I0752619

IMAGES
of America

PACIFIC COAST HIGHWAY IN CALIFORNIA

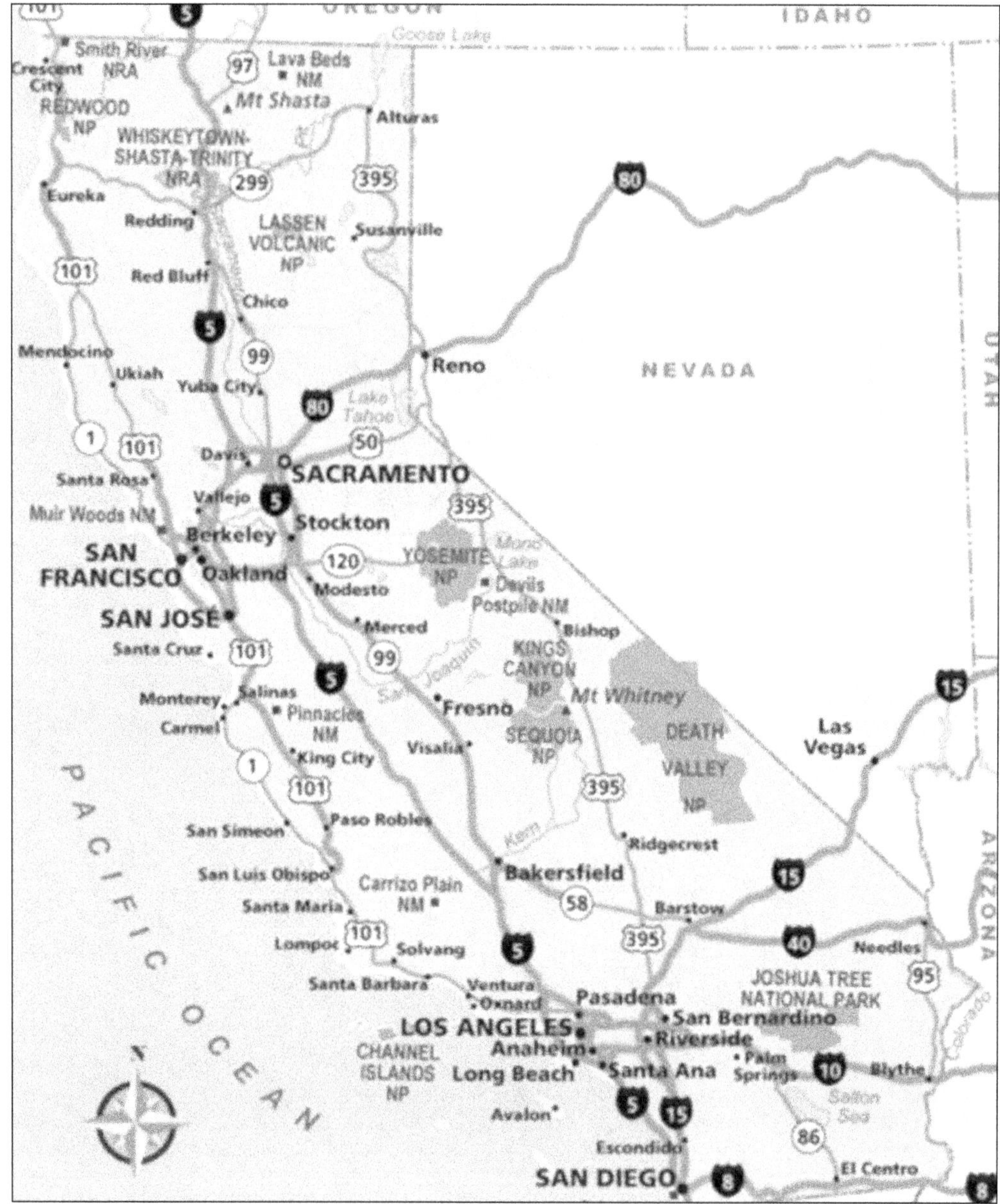

This map of California shows the complete route of the coast highway, which comprises three designated roads. In the southern portion of the state, Interstate 5 becomes the coast highway for 79 miles. Just below San Juan Capistrano in Orange County, Highway 1 becomes the coast highway and follows much of the coast up to Santa Barbara, where it begins to wend along the edge of the coastal mountain range. In Northern California, Highway 101 takes over and winds along some of the state's most scenic shorelines. (Caltrans District 5.)

ON THE COVER: Mugu Rock is one of the most recognizable sights and landmarks along the Pacific Coast. It was once part of the Santa Monica Mountains that was vertically cut and blasted away to create a passage for the coast highway between Malibu and Oxnard. Located in Ventura County, it sits on a jagged shore known for its unpredictable surf, crashing waves, and strong undertow.

IMAGES
of America

PACIFIC COAST HIGHWAY IN CALIFORNIA

Carina Monica Montoya

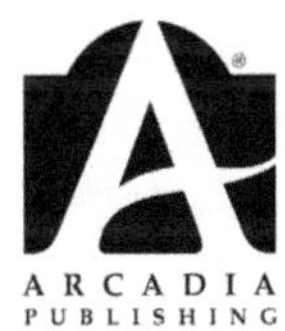

ISBN 9781540228680

Published by Arcadia Publishing
Charleston, South Carolina

Library of Congress Control Number: 2017952569

For all general information, please contact Arcadia Publishing:
Telephone 843-853-2070
Fax 843-853-0044
E-mail sales@arcadiapublishing.com
For customer service and orders:
Toll-Free 1-888-313-2665

Visit us on the Internet at www.arcadiapublishing.com

This book is dedicated to Robert C. Pavlik, who inspired me with his knowledge, invaluable help, and love of California's coast, which he perfectly describes in his poem "El Sur Ranch:" "Brown cows, like dusky storm clouds floating while grazing on a sea green sky, eating the astronomical turf of the Big Sur Coast."

Contents

Foreword

To explore and discover is to be human. From pioneers settling the West to astronauts in outer space and everything in between, these instincts drive our basic existence. In California today, one of the most popular experiences people seek is a journey along the immense and varied coastline via the Pacific Coast Highway.

They can experience adventure traveling through a dramatic landscape at the edge of the continent continuously being shaped by the forces of nature. A viewing platform itself, the highway is seemingly etched into the side of the mountain that falls steeply to the vast expanse of the Pacific Ocean.

They can also discover peace and solitude, as the highway is a means to secure life's necessities in a remote, lightly populated region of the nation's most populous state. One can seek a life away from the trappings of modern civilization or operate a business and rely on a steady supply of visitors to make ends meet.

The California Department of Transportation serves these needs as part of its responsibility to operate and maintain Highway 1. People today expect some things with absolute certainty. Getting where we want to go, much of it on the state highway system, is one of them. But in places where nature still dominates, the reality is closer to what early explorers may have experienced, traversing a rugged landscape from one obstacle to the next.

Perseverance is the theme for living with landslides and shifting shorelines. Highway workers may be kindred spirits of the US Postal Service: "Neither snow nor rain" keep them from their rounds. And yet, massive landslides, shedding rocks and flowing mud, can bring things to a complete halt and disrupt all expectations.

Infrequent but inevitable are heavy storm events, such as those in 1983, 1998, and 2017, challenging the professionals to find solutions that are both expedient and responsible for the safety of the traveling public.

How could we have it any other way? This landscape represents the uncertainty life brings. The Pacific Coast Highway is much more than a feat of engineering; it is a source for exploration and discovery that contributes to our natural human experience.

—Aileen K. Loe
Deputy District Director
Caltrans District 5

Acknowledgments

This book would not have been possible without the support of the California Department of Transportation (Caltrans), particularly Kendra Stoll, senior librarian/history reference at Caltrans Transportation Library and History Center. Her professionalism and diligent work in assisting me by providing reference materials and appropriate images greatly helped in making this book possible. I am also very grateful to Aileen K. Loe, deputy district director of Caltrans District 5, for her contribution in writing the foreword that describes the wonder of a highway that parallels a dramatic coastline.

I want to especially thank Stacia Bannerman, senior title manager at Arcadia Publishing, for her assistance, patience, and understanding during the writing of this book. Many thanks also go to Katie W. Kellett, formerly of Arcadia Publishing. Her diligence in acquiring the necessary licensing agreement for copyrighted material is most appreciated. Thanks as well to Jim Kempert for his professional advice and edits.

I am indebted to Pat Duty, Caltrans transportation engineering technician, and particularly Christopher Mancini, Caltrans transportation engineering technician, for his help in providing me with the majority of images featured in the book. Without his professional and prompt assistance, this book would lack the images that adequately tell the story of the building of California's coast highway.

My deepest appreciation to Dan W. Klos for all his IT help and assistance, and more importantly, his friendship. Thanks also to David E. Cole for his support, encouragement, and help in the early stages of this project; Kenneth D. Bicknell, digital resources librarian/library, archives and records information management at Los Angeles County Metropolitan Transportation Authority; and Matthew Barrett, Transportation Research Library, Archives & Records Management, Information and Technology Services at Los Angeles County Metropolitan Transportation Authority.

I am honored to have had the ongoing professional assistance, guidance, and pre-submission review of this book by Robert C. Pavlik, retired Caltrans environmental planner, historian, and author.

Unless otherwise credited, all images are courtesy of Caltrans District 5.

Introduction

California's Pacific Coast Highway (PCH) is actually an amalgam of three different routes. From Northern California at the Oregon border, US Route 101 passes near Leggett, the northern terminus of the PCH. Highway 1 begins and ends here. It then extends south to Gaviota in Santa Barbara County, where the highway rejoins US Route 101. Several miles south to Ventura County at Oxnard, Highway 1 continues south to San Juan Capistrano, where it merges with Interstate 5 and runs south to San Diego near the Mexico border.

California's coast route was built piecemeal and in various stages. Construction of an early coast road between Carpinteria, a small oceanside city located in southeastern Santa Barbara County, and the Ventura River in Ventura County, occurred in 1911. The road opened a route for motorists to traverse large sections of the coast from San Francisco to Los Angeles. Farther south, a stretch of coast highway in Orange County was built that linked Newport Beach and Laguna Beach in 1913.

Building a road across Arroyo Hondo Canyon, one of the deepest gorges on the Gaviota coast of Santa Barbara County, necessitated the construction of a bridge. The Arroyo Hondo Bridge was constructed as a multiple arch bridge and opened in 1918. The bridge served traffic for more than half a century, but as the years passed, it became too narrow for the increasing traffic and larger vehicles that did not exist at the time it was designed. When US Route 101 was rebuilt through Arroyo Hondo Canyon, it bypassed the bridge. The Arroyo Hondo Canyon Bridge closed, but it was left untouched, perhaps as a symbolic gesture of a great engineering feat by the California Division of Highways. It currently serves as a scenic overlook, a pedestrian walkway, and a vital link in the Pacific Coast Bike Route.

At Point Mugu, a rocky promontory between Malibu and Oxnard, a solid volcanic rock outcrop obstructed the proposed coast road. Initially, a wide path was cut around the rock to facilitate vehicular passage. In 1923, several 30-foot holes were drilled into the rock by workers who scaled the cliff with ropes and filled the holes with blasting powder and 18 tons of surplus World War I hand grenade powder. The blasting produced 108,000 cubic yards of rock, which was used to fill nearby road embankments. By 1924, the 275-foot narrow radius curve around "the Point" gave passage through the area, but the road was later deemed to be too dangerous for motorists, and further removal of rock was required. It was determined that a passage would be cut straight through the rock. The job required 107 tons of explosives and a legion of laborers, along with diesel shovels, bulldozers, and jackhammers to create a 200-foot-deep and 60-foot-wide road passage that took five years to complete. The delays were caused by funding and labor shortages (a not uncommon form of roadblock).

Meanwhile, May Rindge, the widow and landowner of Malibu, refused to allow a road to be built through her Malibu property. Her late husband, Frederick Hastings Rindge, envisioned Malibu to become an "American Rivera" and wanted to keep the property private. May attempted to uphold her husband's wishes by battling with the State of California and the powers of the City of Los Angeles through years of litigation. When her efforts eventually failed, the road was finally built. It was the last coast road section to open linking Santa Barbara to San Diego in 1929.

In 1919, California approved the building of Route 56, which later became Route 1, also called the Carmel–San Simeon Highway. A road along the Big Sur Coast is said to have been first conceived by Dr. John L.D. Roberts (1862–1949). Dr. Roberts lived in Monterey on 160 acres that he purchased from an uncle for $5,000 in 1887. His property later became the city of Seaside, and in 1890, he and his wife Edith (née Maltby) founded the Seaside Post Office. Together, they served as postmasters until 1932. Dr. Roberts also became county supervisor, and served as the local doctor in the rural coast area, making house visits on horseback. On April 21, 1894, tragedy struck on the Monterey coast by Point Sur when the 493-ton steamship *Los Angeles* ran aground, killing and injuring many of its crew. Dr. Roberts "sped" to the shipwreck scene on a two-wheeled, horse-drawn cart that took more than three hours to travel 30 miles. The long trek motivated Dr. Roberts to propose a road that would facilitate overland access to these remote areas along the coast.

Dr. Roberts surveyed the landscape and lobbied for a road along the Big Sur Coast spanning the area from Monterey to San Simeon. He first proposed and promoted the idea of tourist travel along the scenic coastline. State senator Elmer S. Rigdon of Cambria supported Dr. Roberts's endeavors and helped to acquire state appropriation to fund the building of the road. However, World War I necessitated that priority be given only to roads that would serve for military importance. Lobbyists for the coast road amended its proposal of a scenic road to a proposed military road, resulting in the Carmel to San Simeon highway being included as one of six roads of military importance when the Military Highway bill was passed. In 1919, California voters ratified a constitutional amendment that allowed appropriation for the construction of the Roosevelt Highway, named in honor of Pres. Theodore Roosevelt, who died January 6, 1919. The route is also referred to as the Carmel to San Simeon highway, and would later be known as California's Pacific Coast Highway and designated as State Route 1.

Surveying the Big Sur landscape had to be performed on foot and horseback. The construction of a road along the Big Sur Coast began above San Simeon near Piedras Blancas Lighthouse in late 1921. The route would connect Big Sur to Southern California's coastal communities. Big Sur has some of the state's steepest and most unstable terrain. Friable mountains and solid rock had to be carved out and blasted through to build roads and bridges that would be constructed under, above, or alongside unstable hillsides and unpredictable surf. It was not until the mid-1920s when advancements in the production of mechanized equipment, such as steam shovels, enabled modern road building methods to come to the coast.

The designated road proved to be difficult and expensive. One major problem was recruiting and retaining laborers to work in these remote regions. The road project was stopped and put on hold in 1924, just two years into its construction. Lack of funding and a shortage of laborers delayed the building of the road for four years. When construction resumed in 1928, prison labor helped meet the demands of the labor shortage. Prisoners of San Quentin were assigned to labor camps at Little Sur River, Kirk Creek, and Anderson Creek.

Unstable earth along the Big Sur Coast was problematic, and landslides were a continuing problem during construction, but the most challenging work was along the 65-mile section between Spruce Creek and San Simeon. Equipment was lost by falling into the ocean or damaged from falling rock and dirt. Workers were sometimes injured or killed. Eventually, after 13 million cubic yards of rock and dirt were blasted away, the two-lane road was completed and opened in 1937. Because of the tenuous landscape, the potential threat of landslides is always present due to storms, erosion, and earth movement from earthquakes. Since its construction, the repair, reconstruction, and major maintenance of the road have been ongoing.

The building of a highway along the Big Sur Coast was indeed an engineering marvel. It was a learning experience for road builders, to say the least, who first used hand shovels, horsepower, and explosives during the early phase of the construction before more modern mechanized equipment became available. Paving roads also went through a process of trial and error. Concrete was first used as a surface, but traffic weight, earth movement, salt, and erosion proved that a stronger and more durable substance and construction technique was needed. Roads were rebuilt and new roads were strengthened with a grid of steel rebar laid first to reinforce the concrete and then asphalt was laid over it.

The building of Highway 1 along the Big Sur Coast necessitated the construction of several bridges to traverse the rugged landscape. The mountainous coast, stretching between Monterey Bay and San Simeon, consists of many deep gorges and sinuous rivers. Bridges were the only practical solutions in many locations, but getting materials on site presented builders with almost impossible, if not death-defying, circumstances due to the remote locations, unstable ground, and unpredictable weather. One such area, known as Big Creek, was an exceptionally difficult place to build a bridge because of the deep canyon, prevailing west winds, and great distance from any population center. The bridge was built in 1937 as a double-arched reinforced concrete bridge. Its span was 588.9 feet long and 23.6 feet wide, with half-arch side spans on each end. A storm in early 1937 destroyed the wooden falsework used to construct the bridge, and it had to be rebuilt in order for construction to continue. It has two main arch spans measuring 177.5 feet long. Some of the bridges, namely Wildcat Canyon, Rocky Creek, Granite Canyon, Garrapata Creek, and Bixby Creek, also known as the Rainbow Bridge, were also incredible engineering feats. The Bixby Creek Bridge is one of the coast's largest bridges. It majestically stands at 260 feet high and measures 718 feet in length. The construction of the structure required 6,600 cubic yards of concrete, 600,000 pounds of reinforcement steel, and 300,000 feet of Douglas fir wood for its falsework. It opened in 1932. The bridges, particularly Bixby Creek, are some of the most photographed structures to this day.

Devil's Slide, a section along the San Mateo coast, is notorious for landslides. Plans to construct an alternate route away from the slide section with the construction of an inland bypass or tunnel resulted in the construction of the Tom Lantos Tunnels. The tunnels were named after the late congressman Thomas Lantos (1928–2008), who significantly supported and helped acquire the funding for the project. The tunnels opened in 2013 and are two of the longest highway tunnels in California, stretching 4,149 feet in the northbound direction and 4,008 feet southbound.

In central California, the coast road rides along the edge of the mountains and passes over some of the country's most unique and picturesque landscapes and bridges, such as San Francisco's Golden Gate Bridge. The bridge is one of engineering's greatest creations; it is a true modern marvel. Construction of the bridge began in 1933. Despite toiling against nature's obstacles of tides, currents, strong winds, and fog, a 4,200-foot suspension bridge, which was the longest of its kind prior to 1964, linked San Francisco to Marin County over the channel between San Francisco Bay and the Pacific Ocean. It is where the coast highway and US Route 101 meet and cross the strait. Vertical suspenders are attached to two main suspension cables that are slung above the bridge, transmitting the deck weight and load to the two towers erected in the tumultuous bay, as well as anchorages at the bridge's north and south ends.

California's North Coast landscape is largely rural and agricultural, but also abundant in beautiful vistas, ominous cliffs, and pounding surf. Several miles of two-lane coast highway twist, turn, and zigzag, seemingly endlessly, along ocean cliffs and below looming mountains. The North Coast is home to California's coast redwoods (*Sequoia sempervirens*), so it is easy to see how difficult, dangerous, and innovative it was to build the coast roads, tunnels, and bridges alongside and through mountains, and across the country's most beautiful grand forests.

Although the highway running through central and Northern California's coast is prized for its magnificent scenic views, Southern California's Pacific Coast Highway boasts some of the world's most famous places along the highway, such as the Hollywood Malibu Colony, Topanga Canyon and Topanga Beach, Pacific Palisades, the Santa Monica Pier, Venice Beach, Marina del Rey, and the famous surfing spots at Manhattan Beach, Hermosa Beach, and Redondo Beach, to name a few. And the many lighthouses that have stood guard through the years still light up the West Coast.

Almost a century later, California's coast road has endured countless landslides, floods, fires, and earthquakes. New and developing techniques continue to be used to protect and preserve a highway with a lot of history in the coastal communities it serves. Dr. Roberts's vision of a scenic road not only came to span the central coast, but nearly stretches the length of California's entire coastline. Its variably named and numbered route is regarded worldwide as not just a road or a place, but rather a destination, visited by millions of tourists each year.

One

BUILDING CALIFORNIA'S PACIFIC COAST HIGHWAY

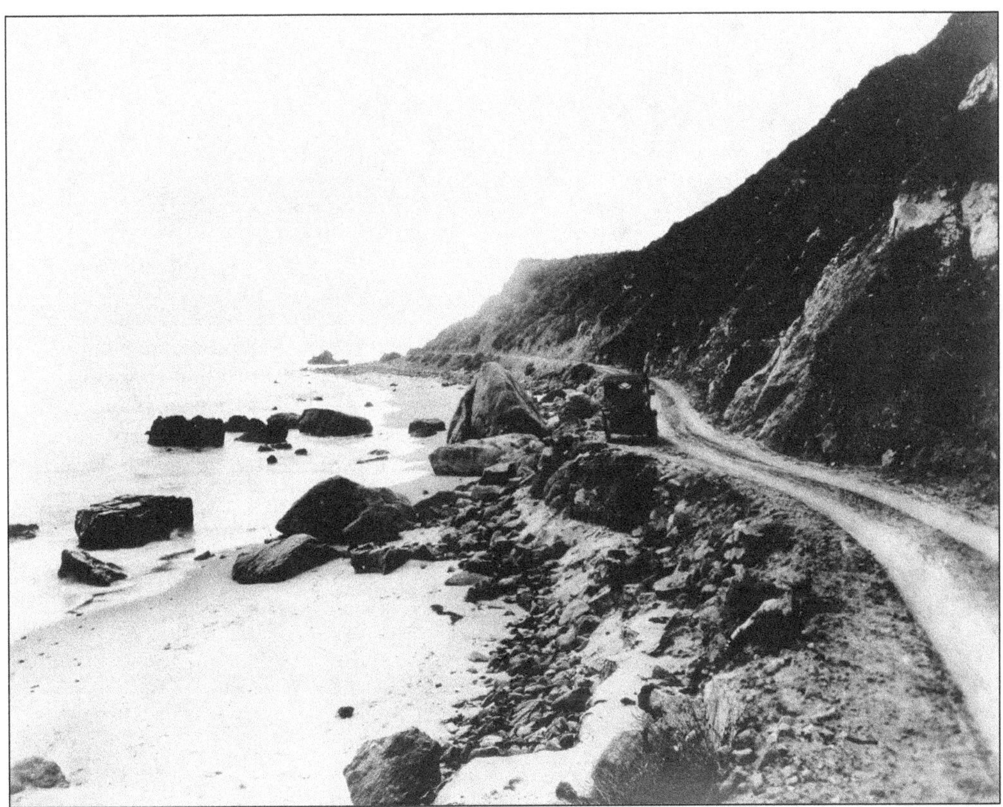

Pictured is a rural road along the Ventura coast looking north in the 1920s. Before the coast highway was constructed, dirt roads along the coast were rugged and dangerous. In some places, travel along the shoreline was the only way to reach remote coast areas. Before automobiles, it took hours to travel a few miles by horse and buggy, only to sometimes find obstruction just around the bend from large fallen rocks. Lower coast roads were often impassable during high tide, storms, and heavy rain. The natural erosion that can be seen along the shoreline, as well as on the hillsides, made for a very unstable road sandwiched between the two. Roads were often washed out or obstructed from mudslides or landslides.

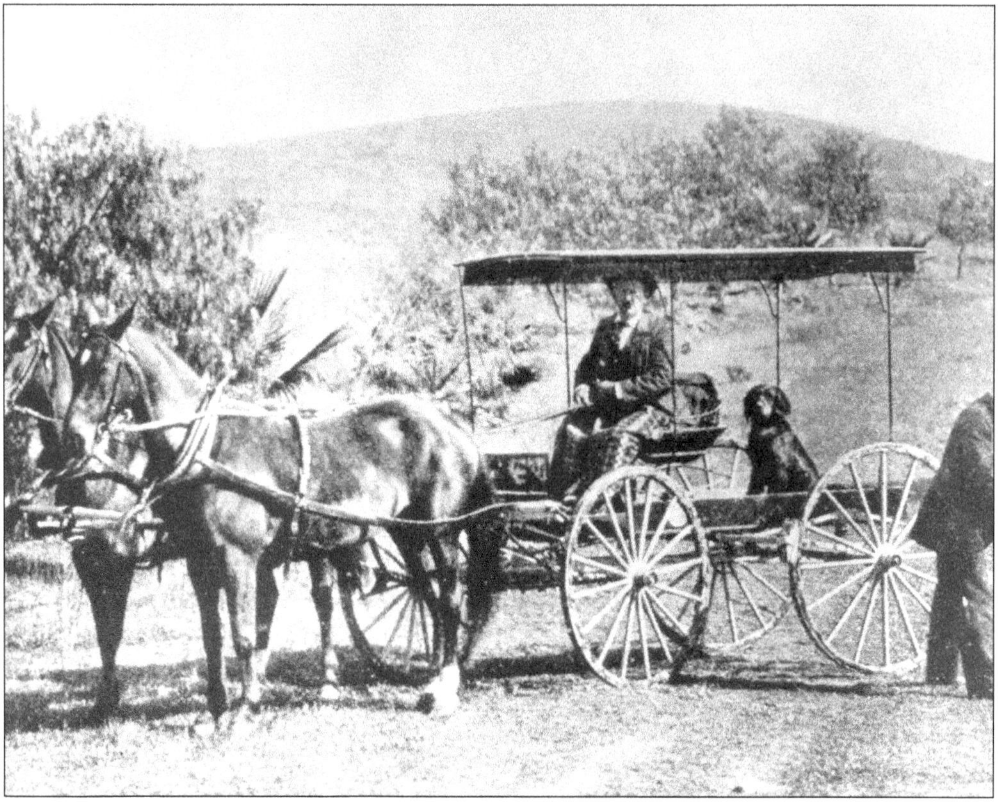

In this early 1900s photograph, a man and his dog sit in a two-horse wagon. Many people believe that the photograph could show Dr. John L.D. Roberts, a rural physician who lived in Seaside and was the first man to envision the need for a coast road. (Copyright California Department of Transportation. All rights reserved.)

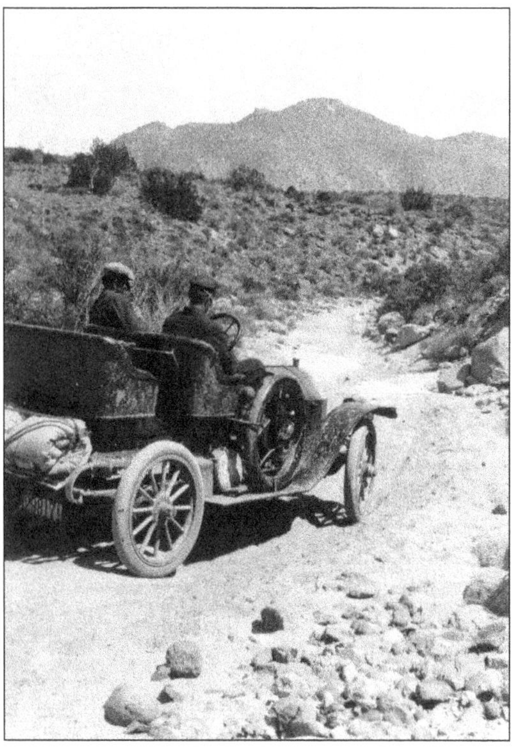

Two men in an open tourer (also called a convertible) drive along a rural dirt road in Monterey. Dirt roads and paths were often washed out during heavy rains and strong storms, which made many communities along the coast remote and isolated from the rest of the state. (Copyright 1910 California Department of Transportation. All rights reserved)

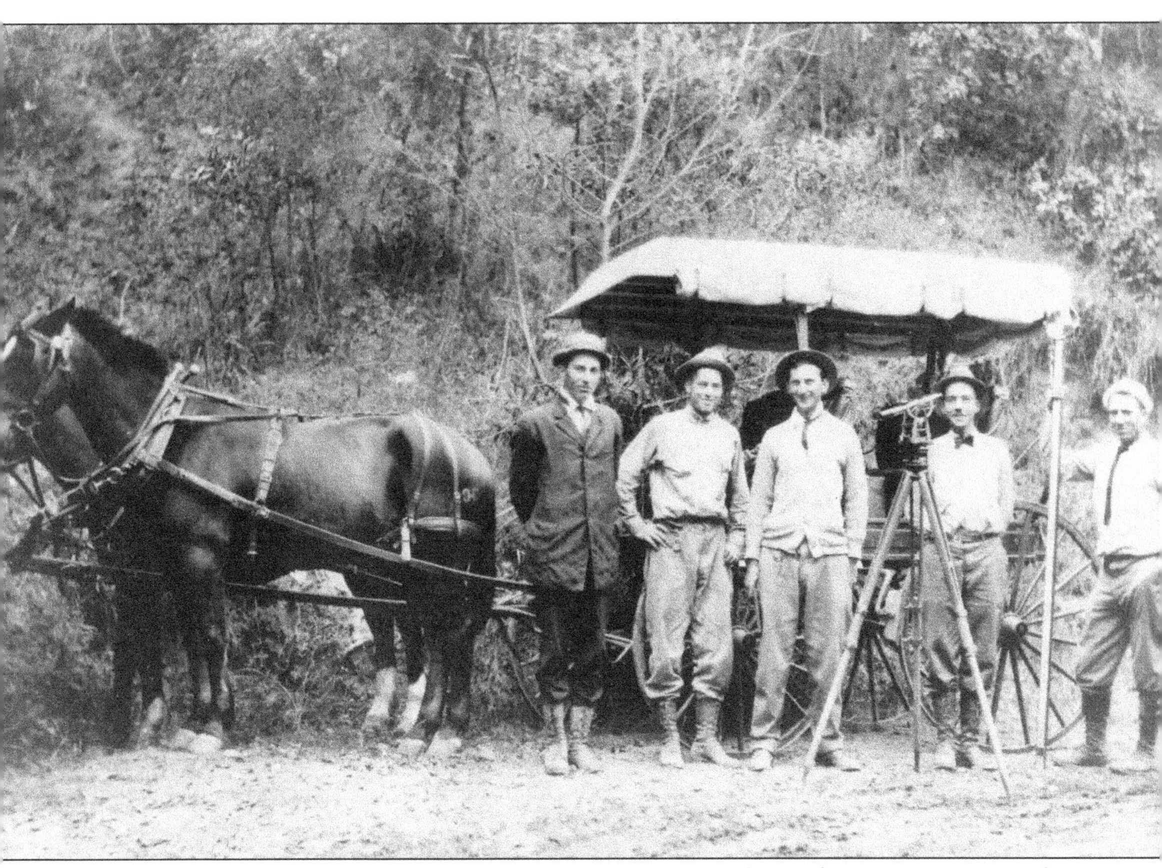

The first official 1911 team of surveyors of the Big Sur Coast is seen here. The survey report stated that the "country between the shore-line and the coast range of mountains running parallel with the shore-line from San Carpoforo to Point Sur is probably the roughest piece of coast-line on the whole Pacific coast."

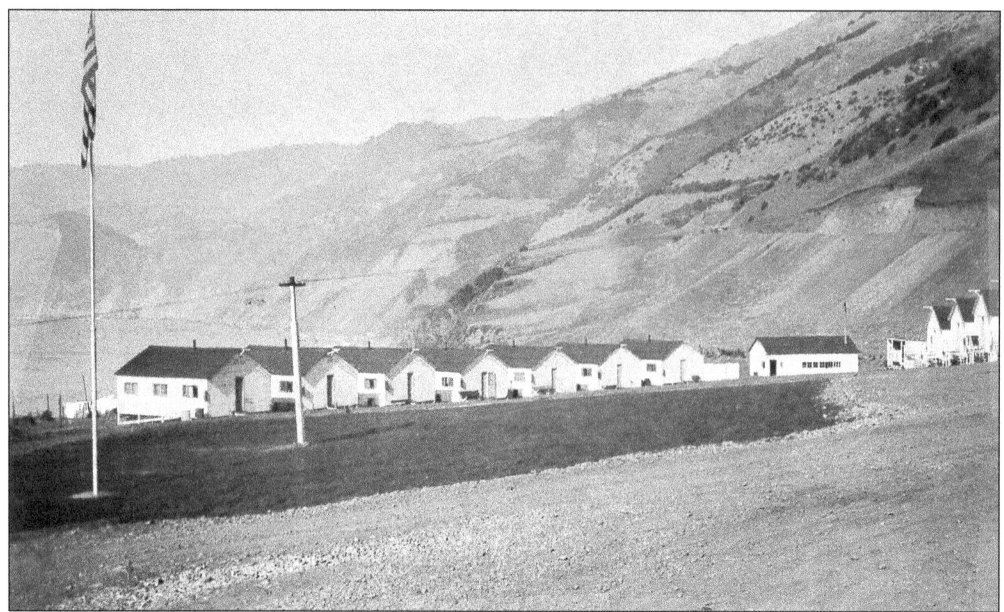

In 1915, a law was enacted allowing the use of convict labor on state highway construction. The benefits of convict labor were threefold: It filled a need for labor, relieved overcrowding in prisons, and was a means of constructive rehabilitation. Temporary camps were first established at Salmon Creek and the South Fork of Little Sur River. As highway building progressed along the coast, camps were relocated and additional ones were established. Pictured is Camp 29 convict labor quarters at Kirk Creek in 1932. This was the first convict labor camp initially established at Salmon Creek in 1928, until the highway to Big Creek was completed. It was then moved to Willow Creek, and later to Kirk Creek. The camp accommodated 120 convicts and 20 free men. Below is a portion of the camp when it was at Willow Creek.

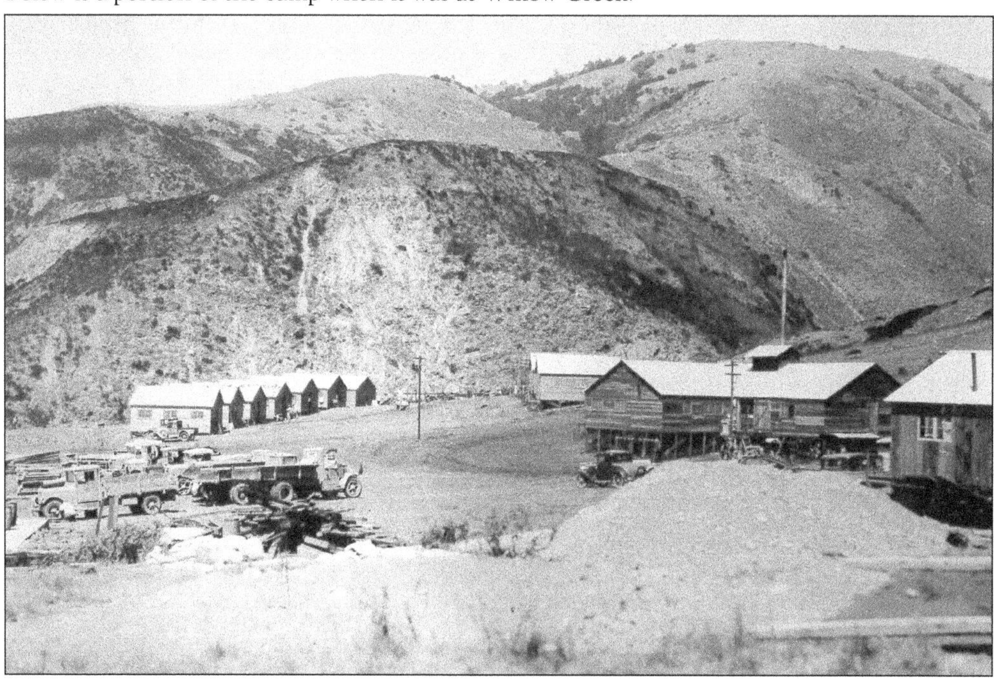

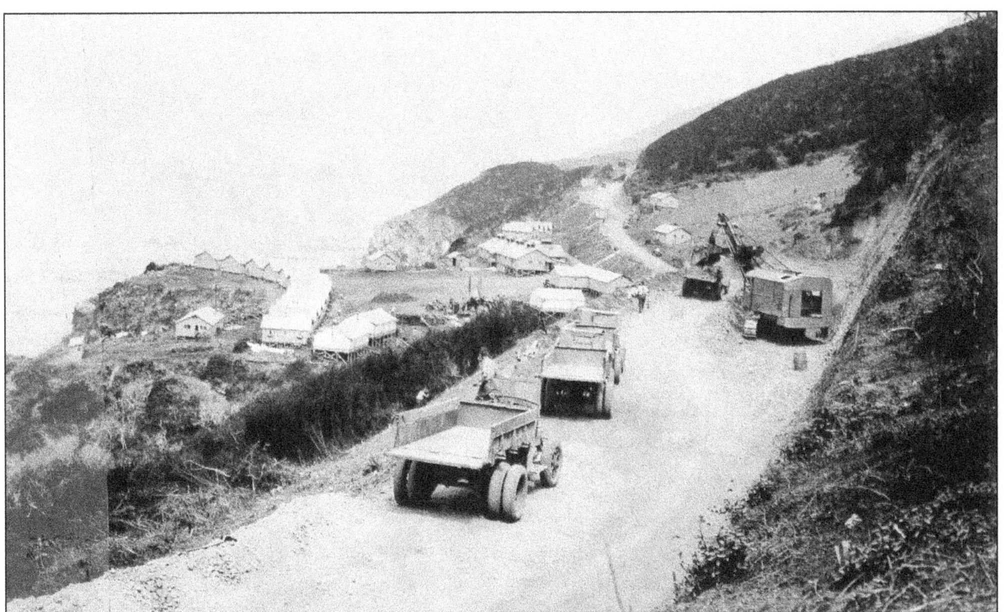

Pictured is the Anderson Canyon Labor Camp located along the Big Sur Coast. It was the largest convict camp, established in 1932. It remained there until the completion of road construction along the Big Sur Coast. After construction of the road south to Big Creek was completed, work began on the reconstruction and realignment of a 10-mile section of road between Anderson Canyon and Big Sur, which was first built in 1924. As pictured at right, free men were assigned skilled labor tasks that included the operation of motorized equipment, and were responsible for delegating and supervising the convicts' work.

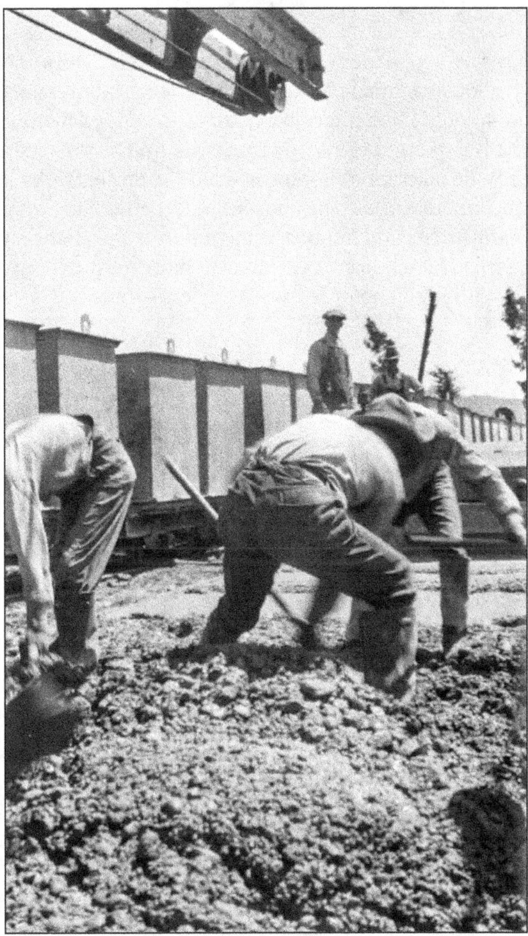

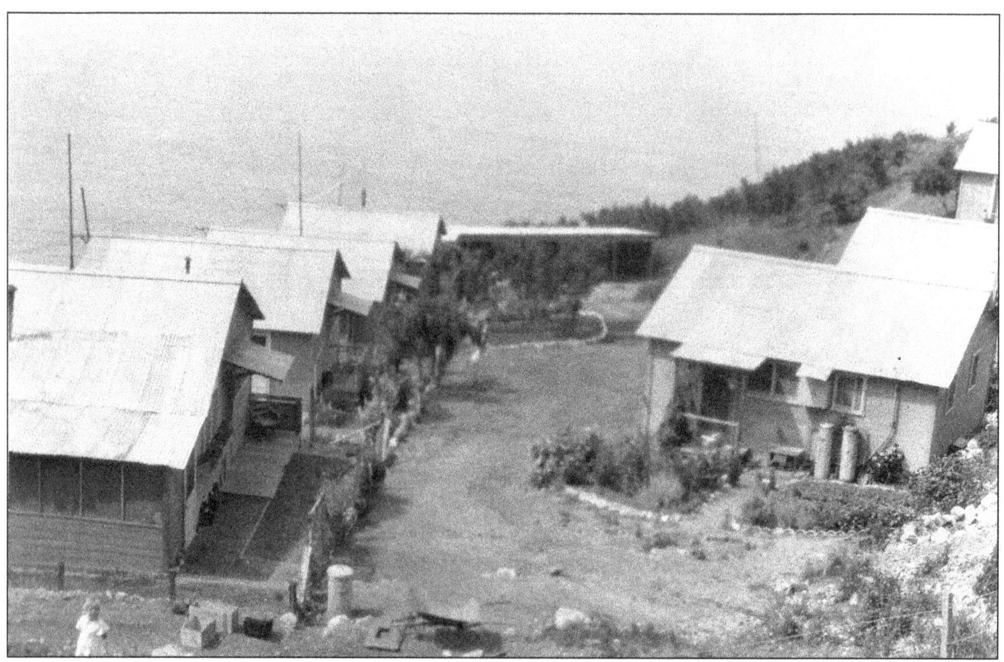

This 1931 photograph shows Camp 26 in Monterey along the Big Sur Coast. The remoteness of this location made it difficult to attract laborers because projects took months and often years to complete. The temporary quarters provided adequate space, but it was still difficult living under the harsh conditions of wind, rain, and storms for lengthy periods of time. Convict labor helped provide additional bodies needed for projects. As seen below, small bungalows were provided for free laborers and their families. After the roads were built and the workers vacated the bungalows, many artists and musicians moved into the abandoned structures. The remoteness of the area, cast against a backdrop of ocean and picturesque mountains, provided them a secluded environment that allowed their creativity to flow—plus, it was free.

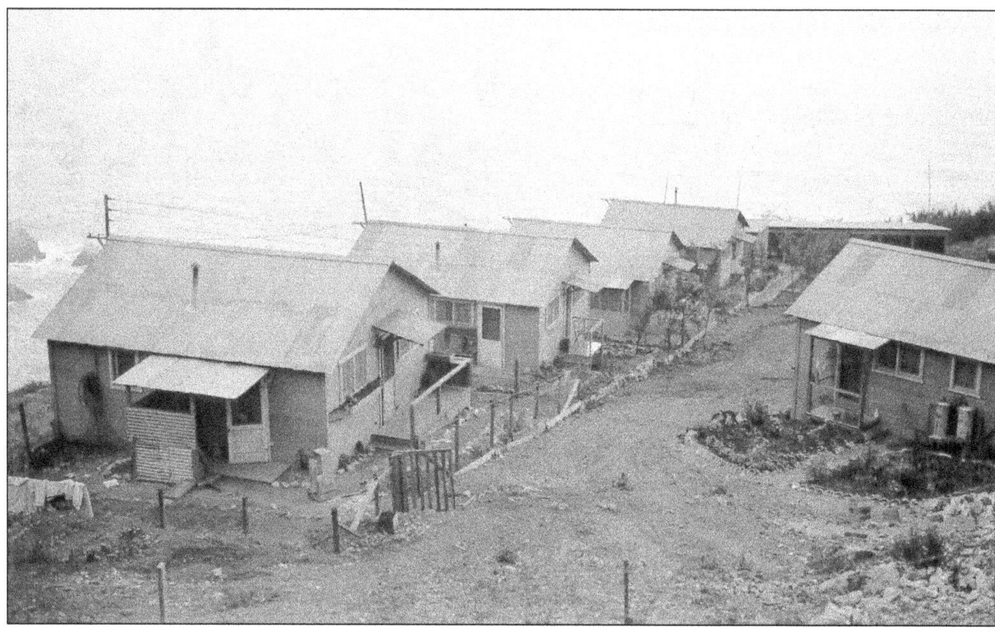

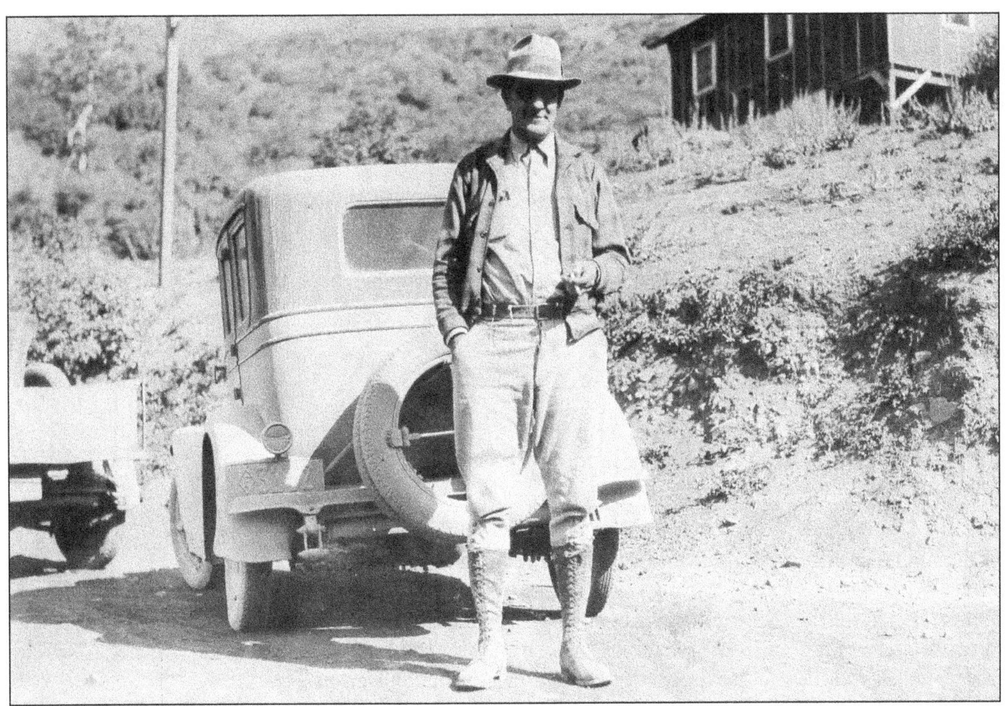

Superintendent B.H. Henry is pictured above in Monterey at a labor camp in the 1930s. Superintendents oversaw building projects, but in the early years of convict labor, superintendents and engineers were not trained to supervise convicts, resulting in inefficient use of labor. However, after reorganization of the Department of Public Works in 1928, the Division of Prison Road Camps was assigned under its Division of Highway Construction, where trained prison guards provided supervision. Below is a foreman's cabin in a free labor camp in 1932. It was a charming location, offering the chance to live in a place nestled in the mountains with an ocean view. Many families often lived in the quarters due to the length of time it took to complete road projects. The occupants made their cabins and landscape around them look inviting, comfortable, and homey in an otherwise remote and desolate area.

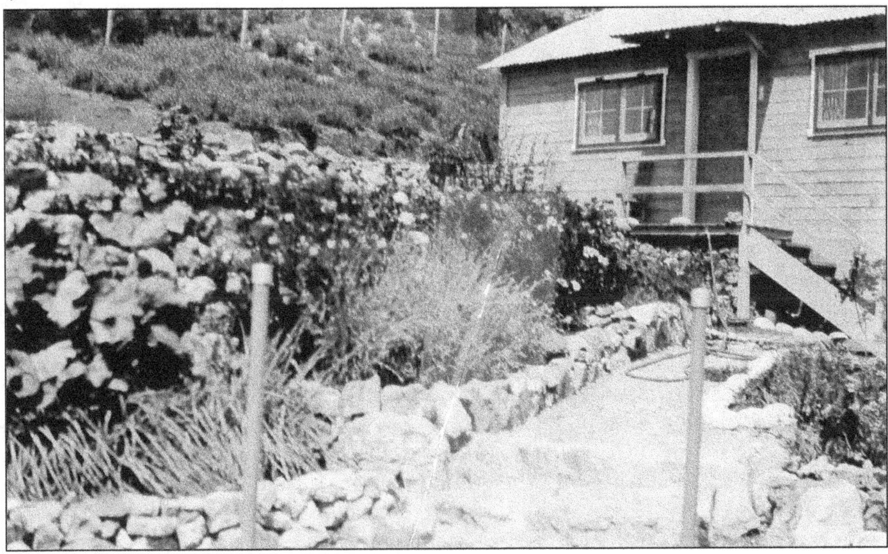

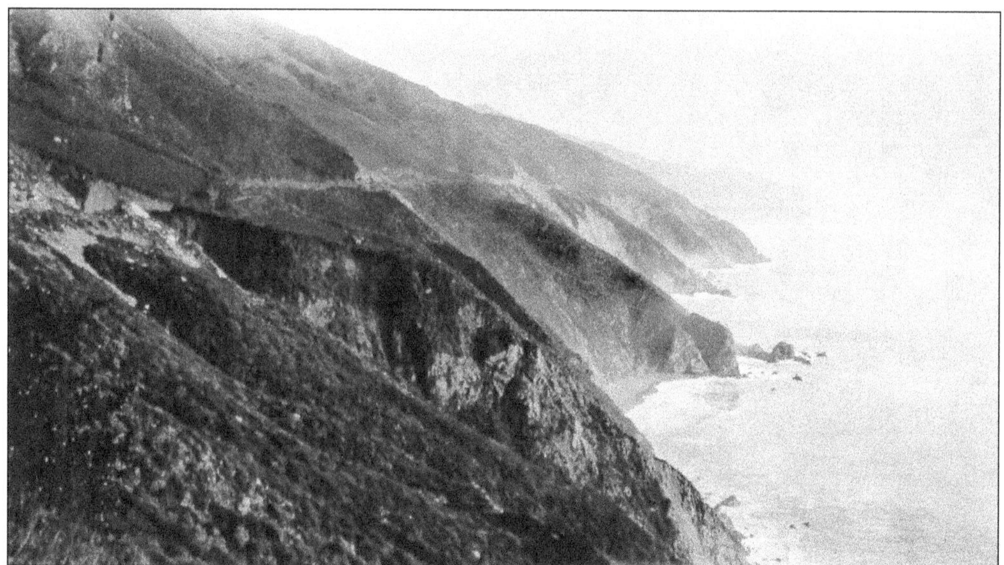

This photograph of Lafler Canyon was taken in the late 1920s. Located in the Big Sur Coast region, it is actually a valley at an elevation of 112 feet above sea level. The area was one of the most isolated and seemingly precarious sites chosen for road construction. The coast road would later be built to follow close along the edge of mountainsides and steep cliffs.

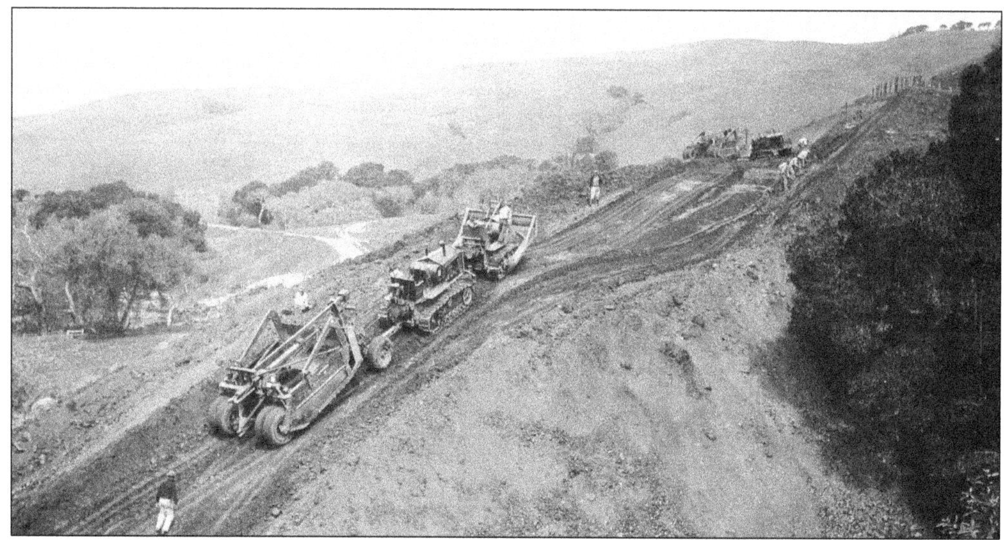

Grading operations along the Monterey coast are seen here in the 1930s. Grading prepares the surface of the road by removing irregularities until a desired smoothness is achieved. Irregularities are removed by the grader cutting into the ground surface. It then drives back and forth across the road, sometimes adding fillers to the surface as needed. Grading on a cliff is dangerous for the operators due to the potential for the heavy equipment to roll over.

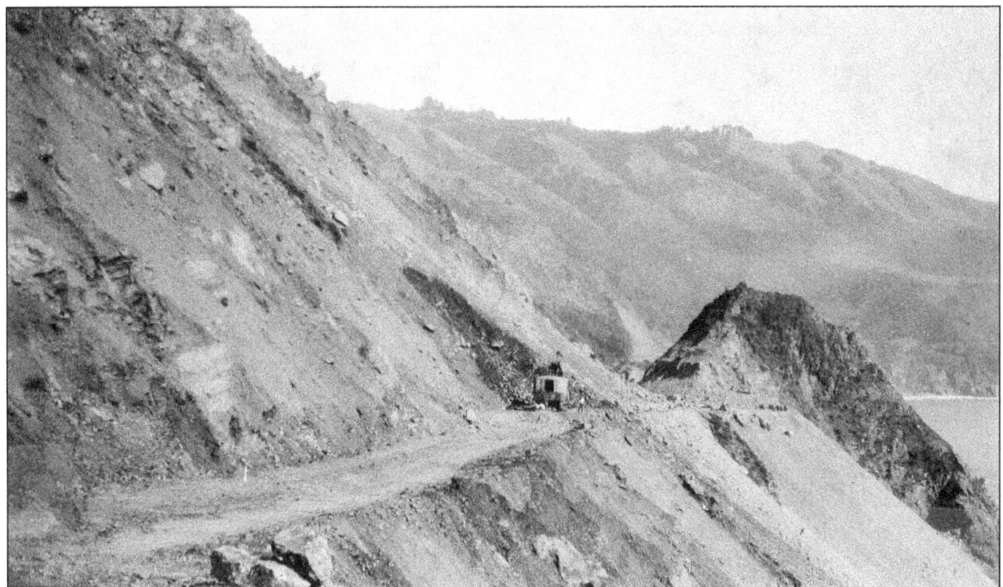

Pictured is a 1934 landslide at Big Creek. Excavation can produce a landslide and weaken ground. The potential of landslides along the Big Sur Coast is always present due to the natural relationship of geology and geography, which can lead to ripe conditions to trigger landslides. Natural coastal erosion, earth movement from earthquakes, winds, and excavations make a landslide more likely to occur.

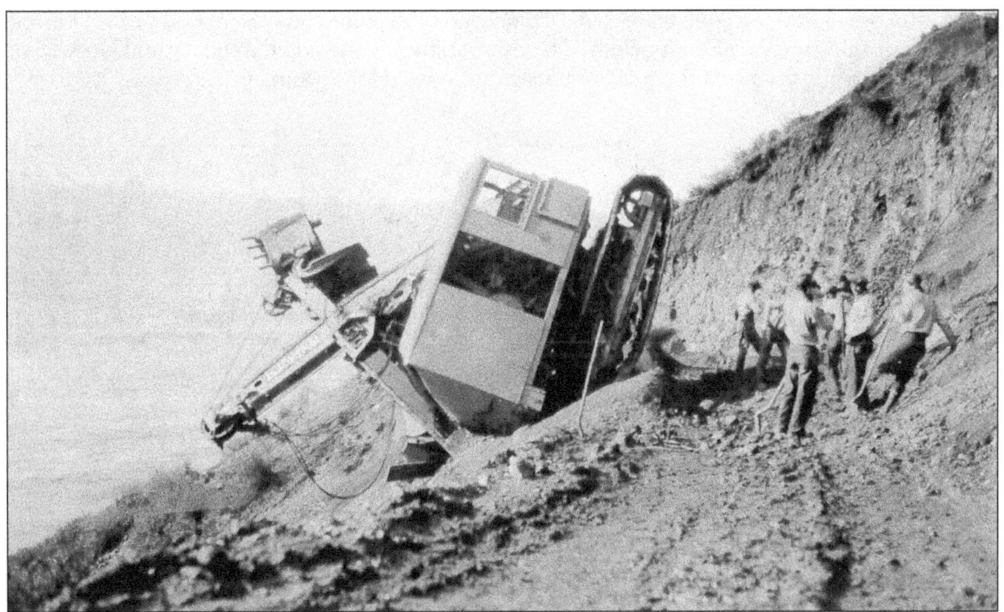

In 1932, a power shovel rolled over while removing dirt from a hillside on the Monterey coast. It stopped short of falling into the ocean. Excavating is considered to be one of the most dangerous and hazardous construction operations for both builders and heavy equipment operators because of the very high potential for injury and for damage to, or total loss of, the equipment.

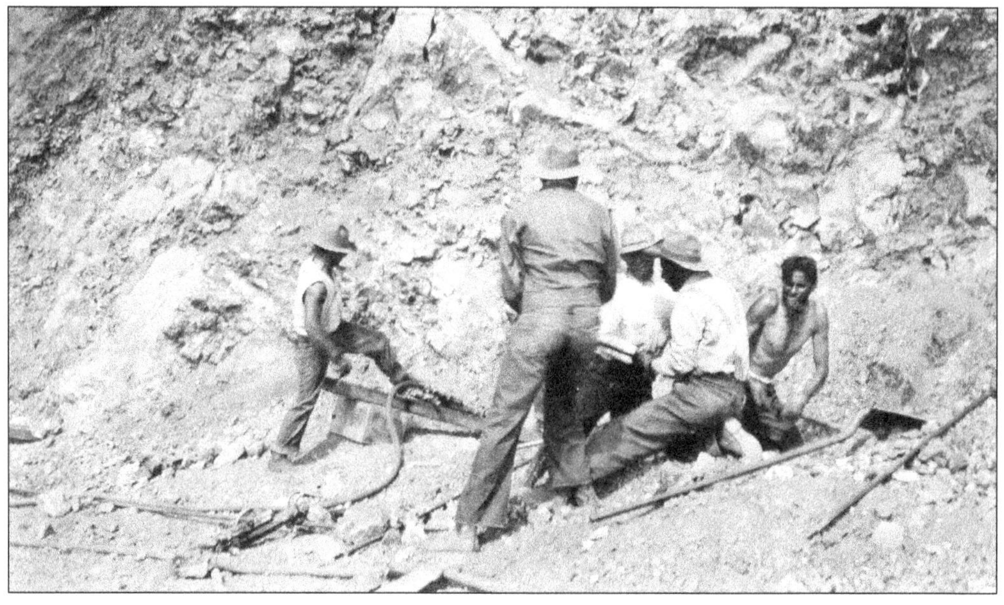

Road construction laborers are pictured above excavating the hillside at Limekiln Point using hand picks, shovels, and steam-powered drills. Both human and mechanical power were necessary to break hard rock and move tons of earth. Hand- and machine-powered drills were used to bore holes into the mountain for blasting. Below, a worker loads dynamite into a mountain hole at Limekiln Point, also referred to as gopher holes. It was reported in 1939 by the Federal Writers' Project of the Works Progress Administration that "at the curve around Limekiln Point, 163,000 yards of rock had to be excavated in 1,000 feet; one blast of 70,000 pounds of dynamite moved 95,000 yards, blowing 75,000 yards into the sea 300 feet below." The excavation not only altered the natural landscape, it also permanently fractured the rock, resulting in the need for ongoing maintenance.

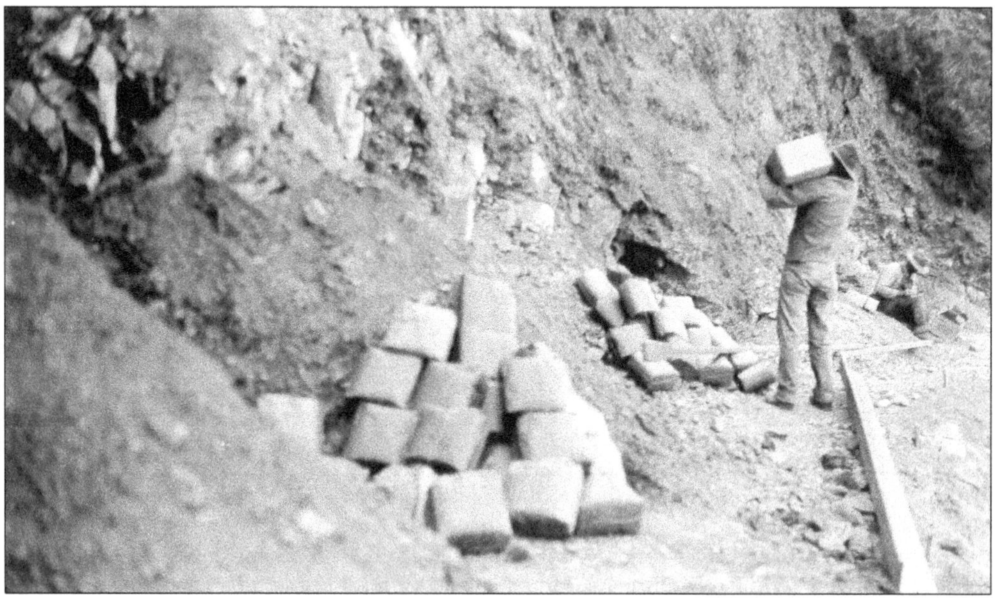

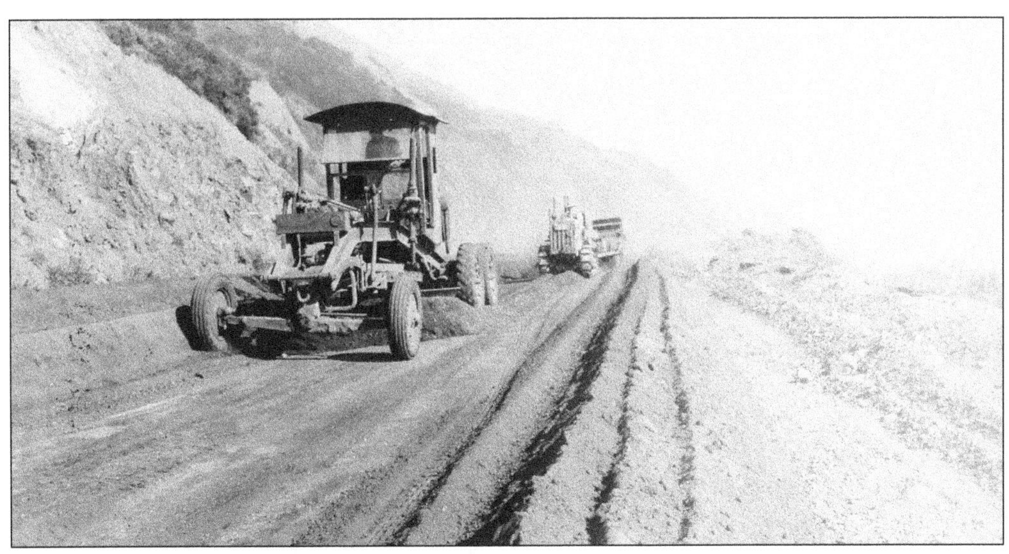

Motor graders are seen above in 1932 along the Ventura County coast. A grader is equipped with a long blade that moves earth to create a flat surface on dirt or gravel roads. In building the coast highway, motor graders were used in the construction of paved roads by preparing the base to create a wide, flat surface where asphalt would be laid. Graders typically have three axles with the cab located above the rear axle at the back end, the engine and third axle at the front, and a long blade between the front two axles. Below is an early model grader with two axles. Its cutting blades can be seen underneath between its two axles. This type of grader was primarily used for cutting header footings.

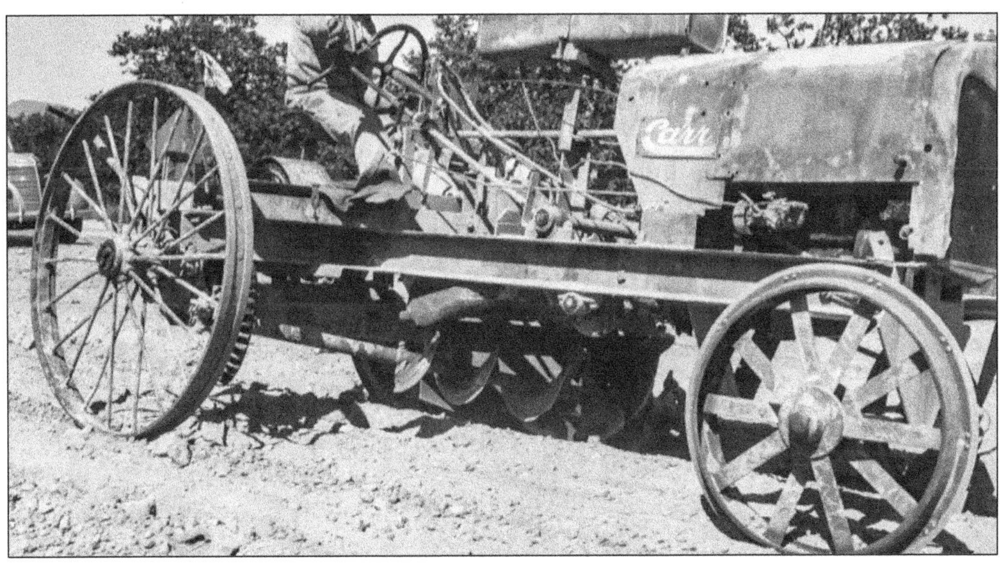

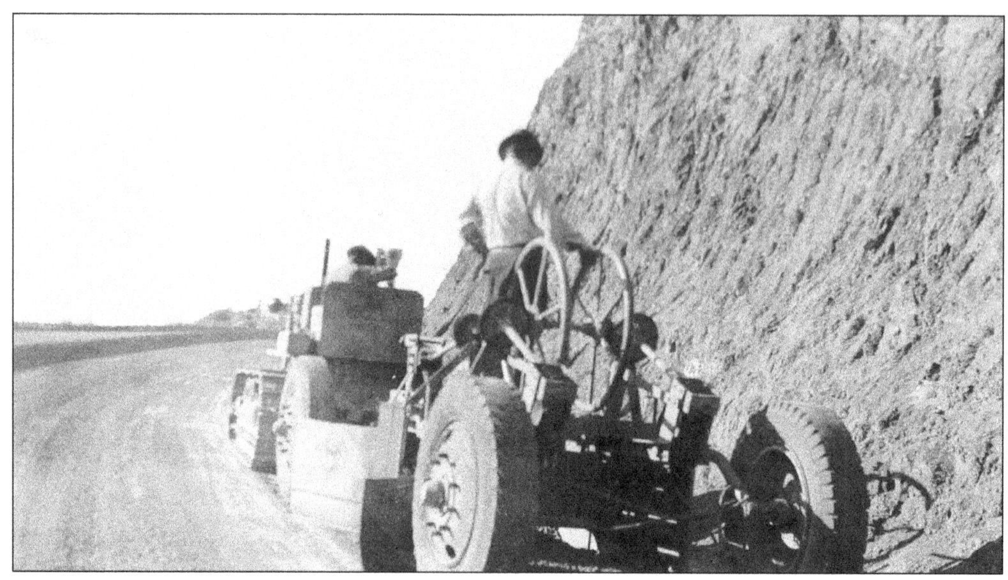

Above, shoulder grading is in progress along the Northern California coast in 1939. Shoulder grading requires road grading material to be deposited along the shoulder to provide a boundary between the road and a hillside. The road shoulder provides a stable surface for vehicles to stop, and also serves as a drainage and small debris catching area away from the road. However, unprotected hillsides are still subject to landslides at any time. Road shoulders below unprotected hillsides are often littered with fallen rocks and debris. Below, two road graders create a flat surface of loose gravel along a stretch of highway in Ventura County in 1947.

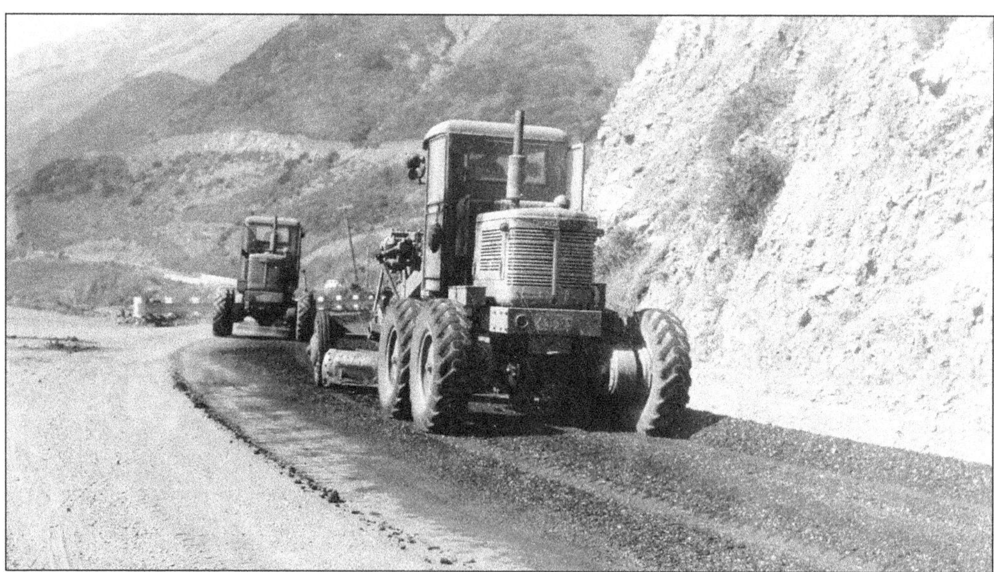

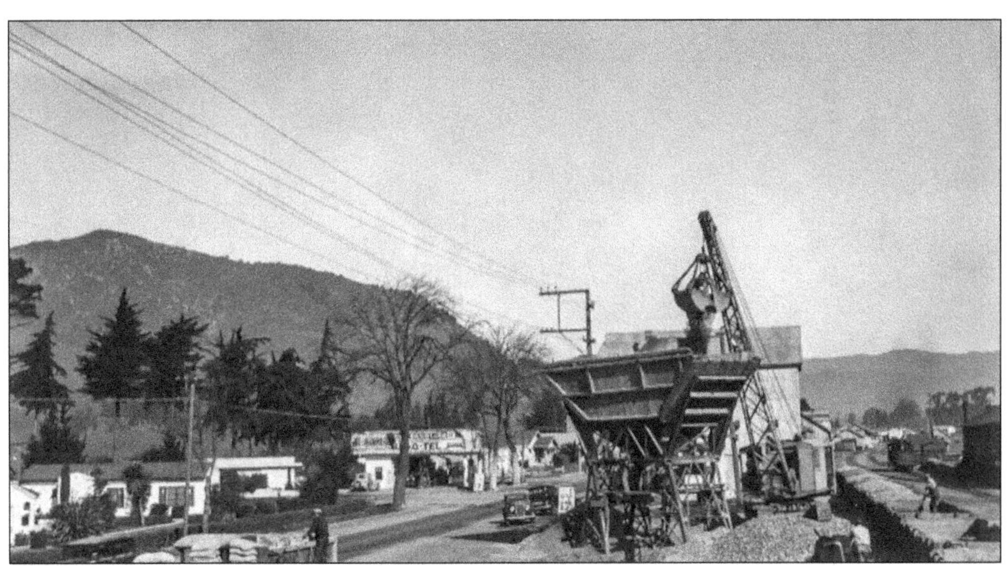

An onsite portable batching plant is seen above along Highway 1 in San Luis Obispo in 1934. The plant mixes various ingredients to form concrete. The batching plant is equipped with a silo, aggregate bin, and an aggregate conveyor. Portable batching plants were used for temporary site projects and provided a steady supply of asphalt without any delays from a supplier or delivery trucks. It allowed projects to be completed sooner, and labor to be used more efficiently. After a project was completed, the batching plant was moved to the next site. Below, workers manually load thousands of pounds of cement into a hopper.

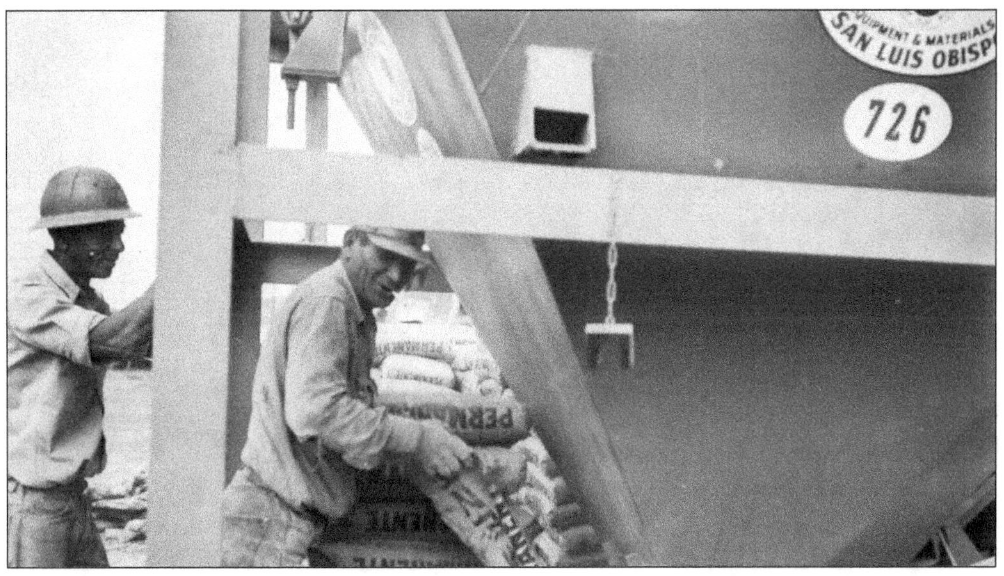

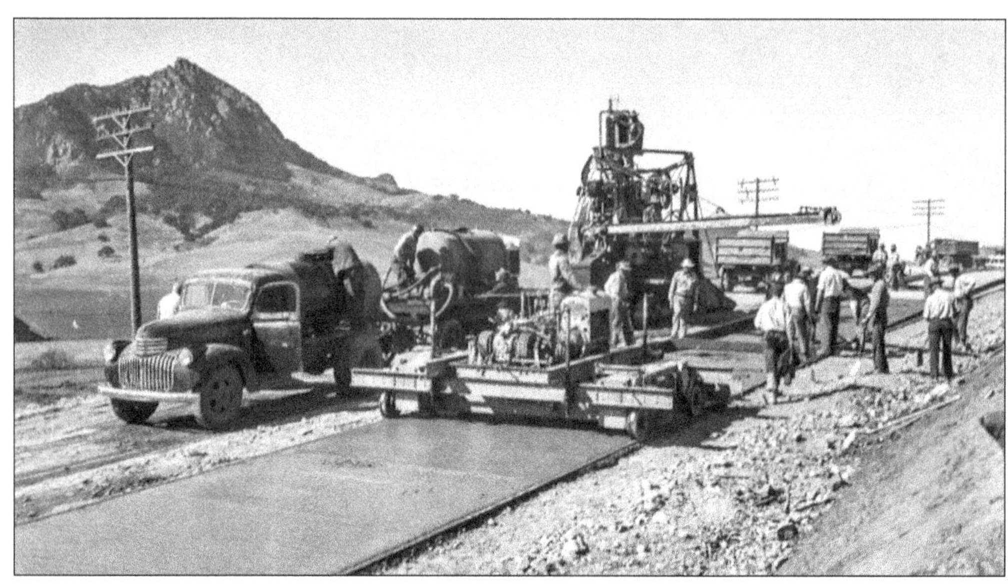

A paving crew in action is seen above in San Luis Obispo in 1941. In the foreground is a paver, and in the background are dump trucks that supplied the paver with hot asphalt and were able to discharge concrete faster than truck mixers. Dump trucks were loaded with cement at an asphalt plant and hauled to the work site. Asphalt is typically applied in layers up to several inches thick for highways and is usually produced at an asphalt or cement batching plant, unless a portable plant is used. Below is another view of a paving crew laying asphalt. A continual flow of paving material is crucial when paving commences. Here, an on-site batching plant produces the asphalt. After the asphalt is laid, rollers behind the paver follow and compact the asphalt.

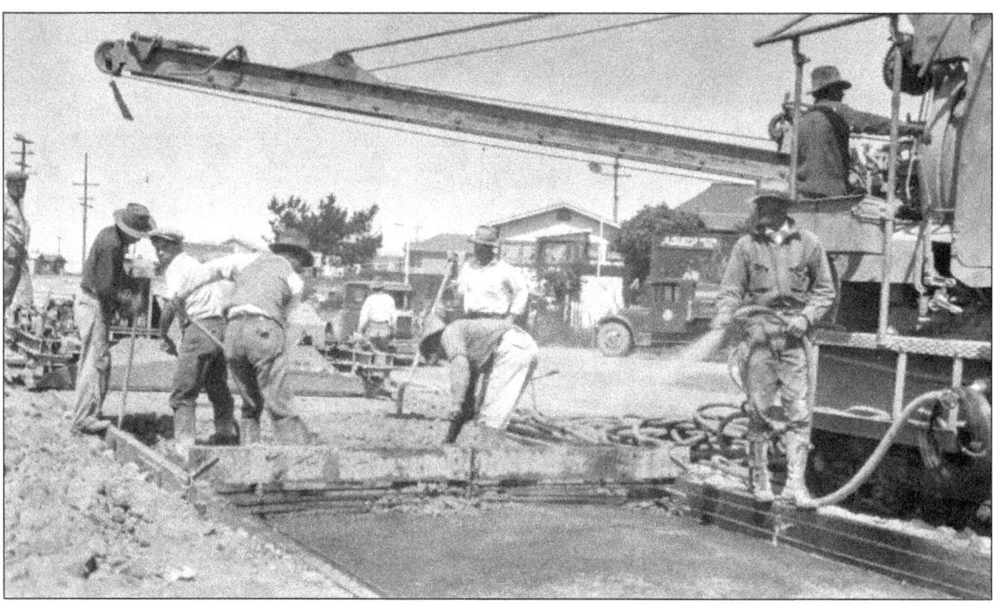

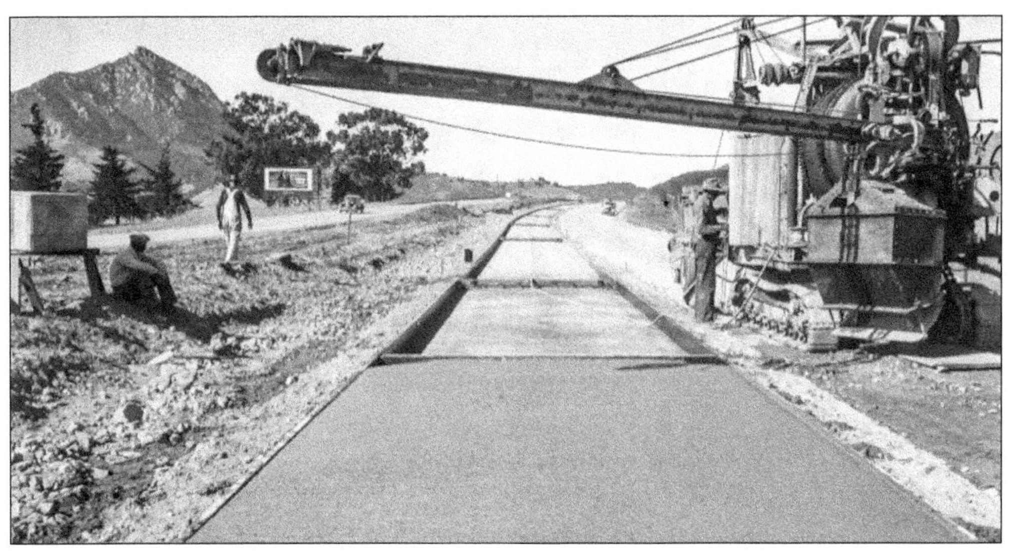

Pictured above is a completed section of pavement subgrade on a section of highway in San Luis Obispo in 1941. Weight from vehicle traffic is transmitted through the pavement to the subgrade, and the weight on the surface of the pavement is then transmitted through the structure, which can destroy the integrity of the subgrade. In the early years of road design and construction, the heavy traffic and larger vehicles of today were not anticipated. Below, workers are seen pulling a diagonal float by hand after the asphalt was laid.

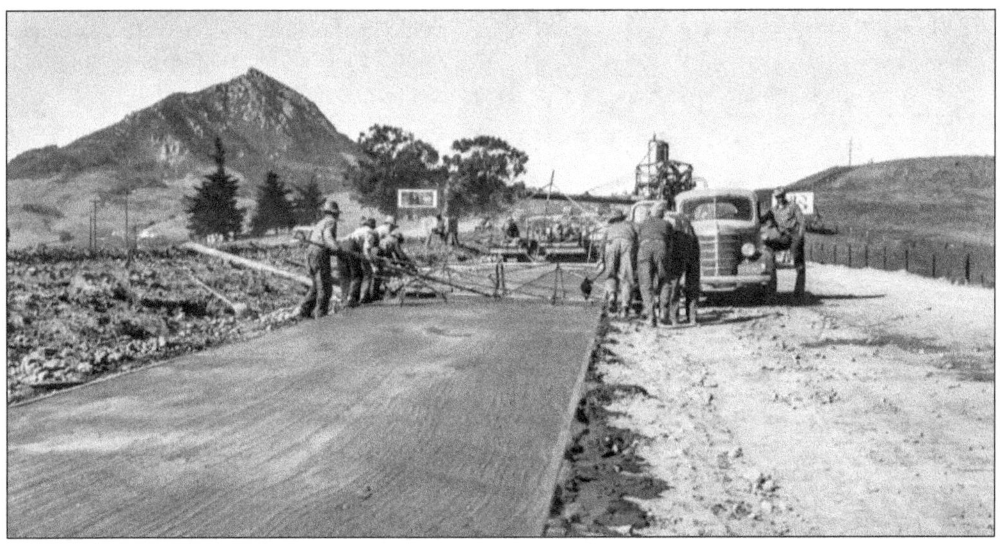

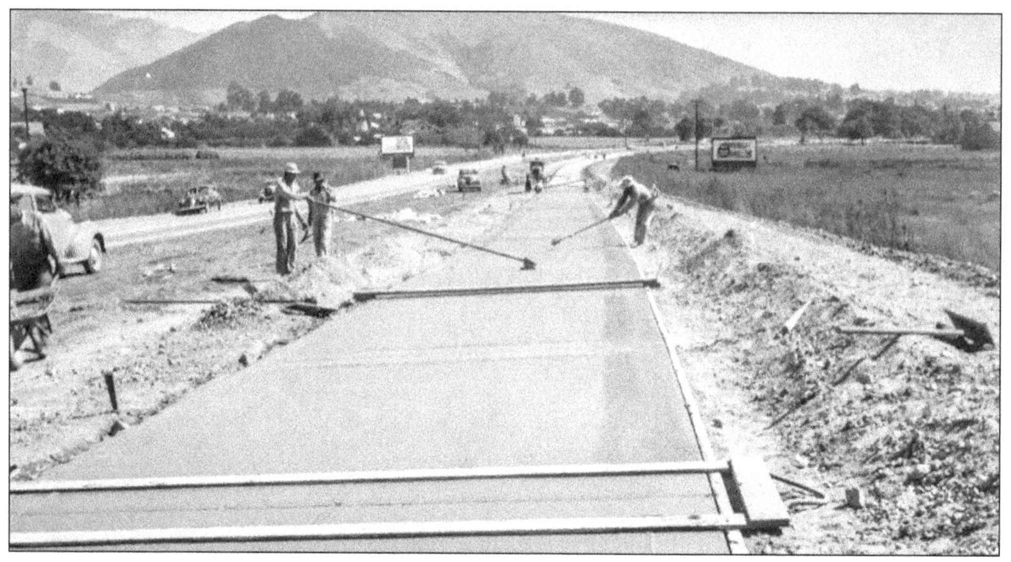

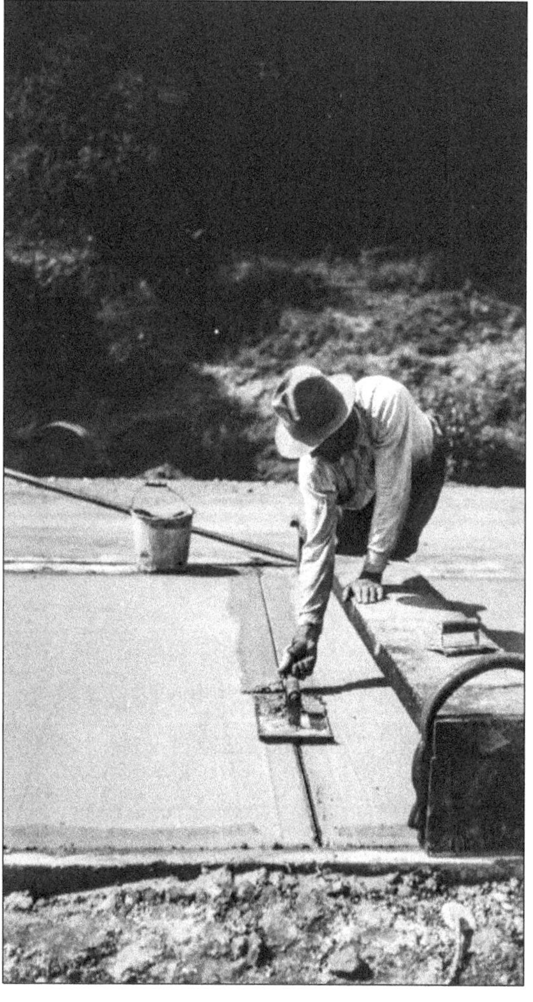

The workers above are finishing joints on the San Luis Obispo section of highway. Most asphalt pavements contain longitudinal construction joints because paving long sections in one pass is not possible. The strength of the pavement largely depends on its joint performance. Most jointed concrete pavement failures are found at the joints. For coastal environments, an expansion joint placed at specific locations allows the pavement to expand without damaging the pavement or other sections of nearby pavement. At left, a worker uses a double edger tool along a road seam. The edger produces a radius at the edge of a concrete slab that smooths the edges and prevents chipping.

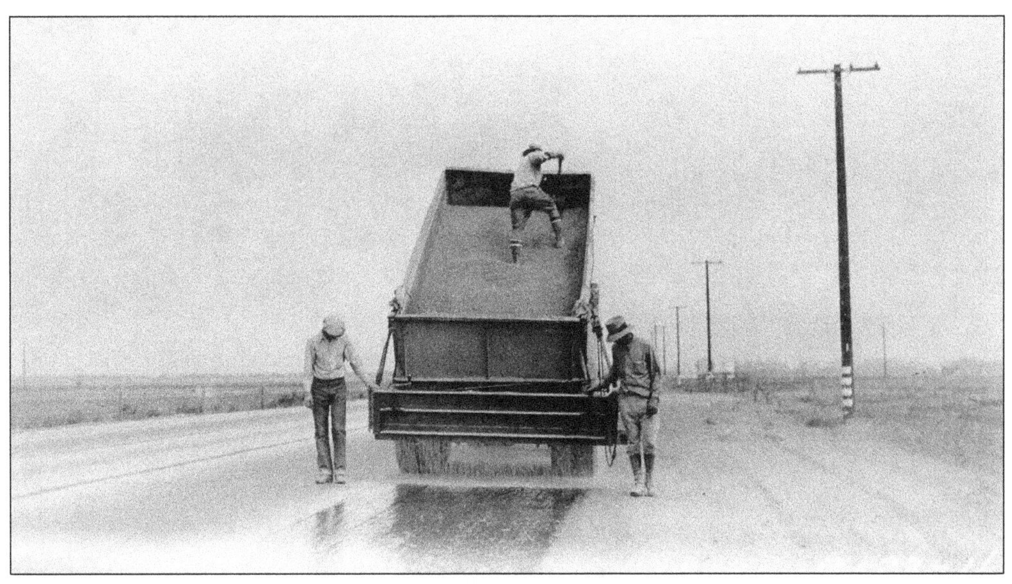

Workers are seen here along Highway 1 in Southern California in 1931. Above, a worker is shoveling gravel from the open-box bed of a truck after a layer of liquid has been sprayed on the surface. While the truck slowly drives in reverse, two workers on each side ensure the evenness of the laid gravel. Below, one man guides a screed from the front of the truck and ensures that excess wet concrete is cut off in order to bring the top surface of the slab to the desired grade and smoothness.

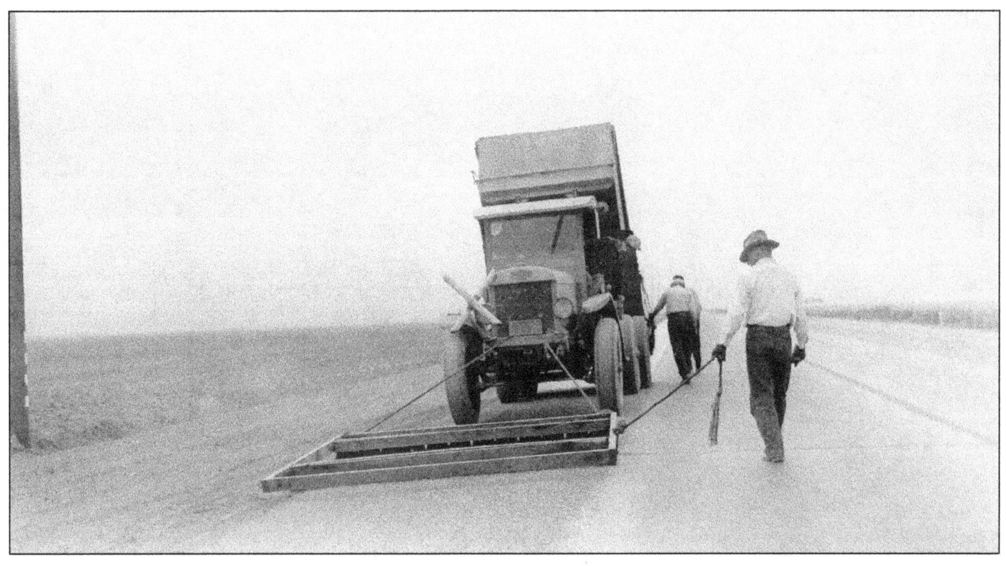

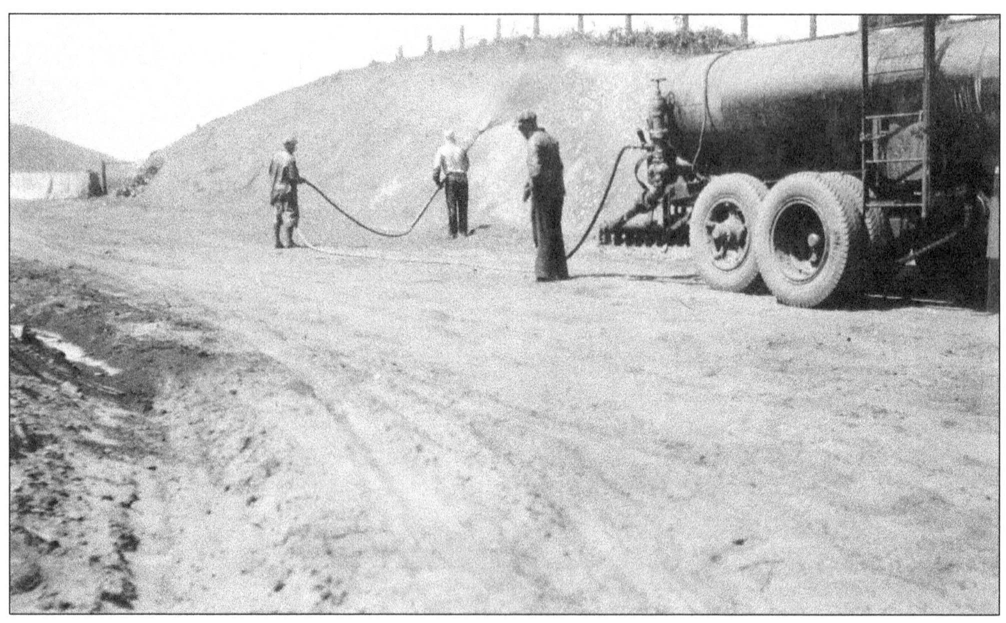

Road workers in 1932 spray a small hillside along the coast road to control erosion. Used motor oil was once used to spray on roads and small hillsides to help repel water, but it has since been banned for environmental purposes. Modern methods attempt to control the soil by using organic compositions such as mulch, compost, or manure. Soil can better absorb water if it is high in organic material. Below, workers install rock riprap on a slope for erosion control. Riprap is placed in a stair-step fashion, anchored with rebar, and then back-filled with dirt to form a solid bank.

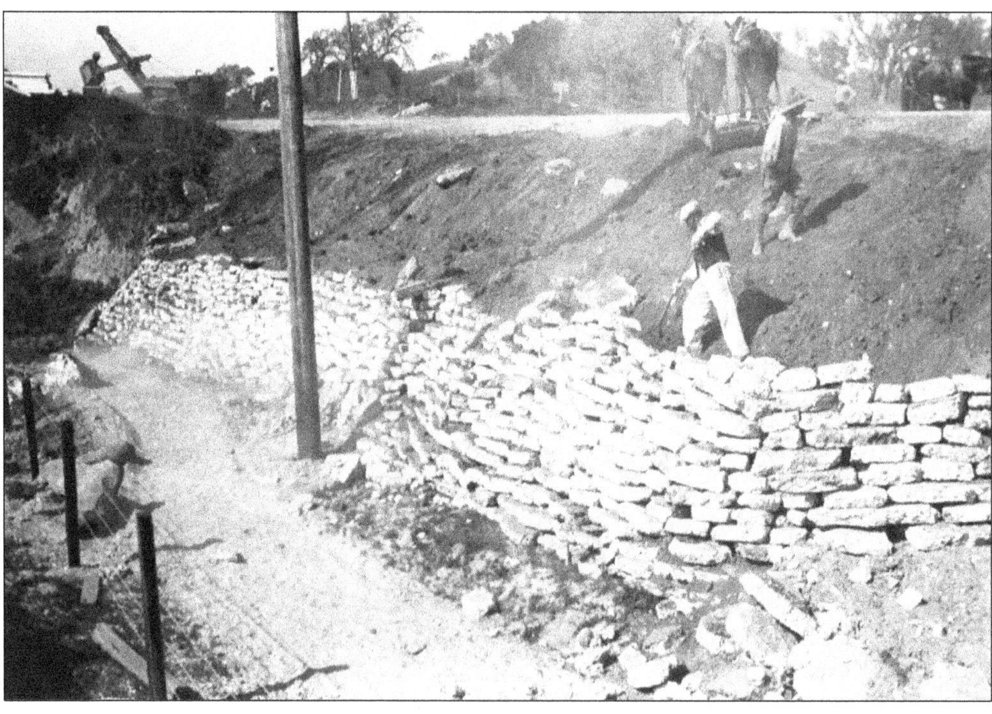

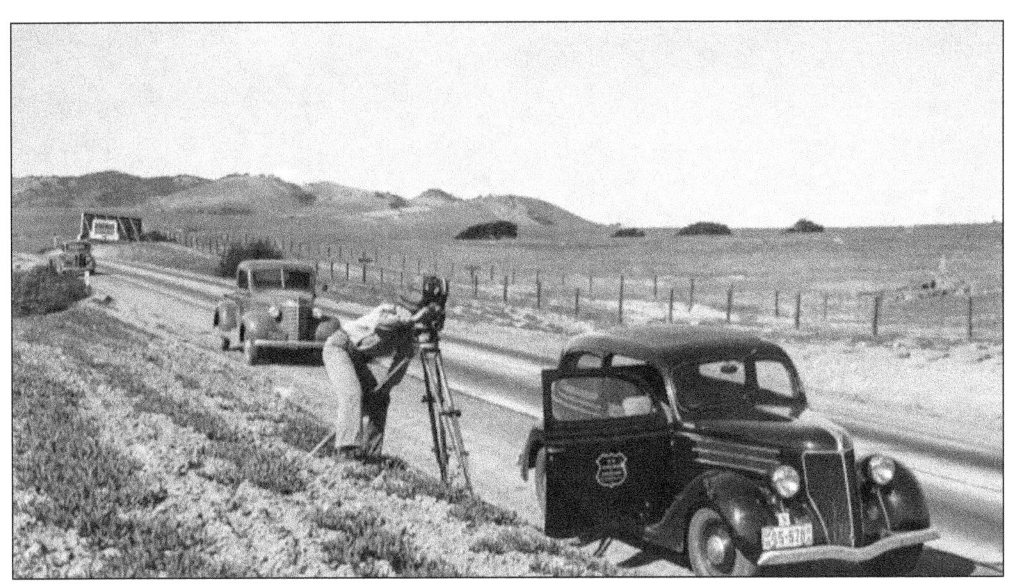

Above, in 1939, a US Department of Agriculture employee films the ice plant landscape to show erosion control along the coast highway in San Luis Obispo. Ice plants were also used in the neighboring counties of Ventura and Santa Barbara. Below, a section of ice plants along San Luis Obispo's coast highway was infected by spittlebugs and had to be eradicated by burning. Although spittlebugs can be found throughout the United States and on most plants, the bugs thrive on plant juices. The infested plants had to be removed and new foliage replanted. Ice plants are ideal for coastal environments due to their ability to tolerate harsh winters, drying winds, salt, and scorching summers.

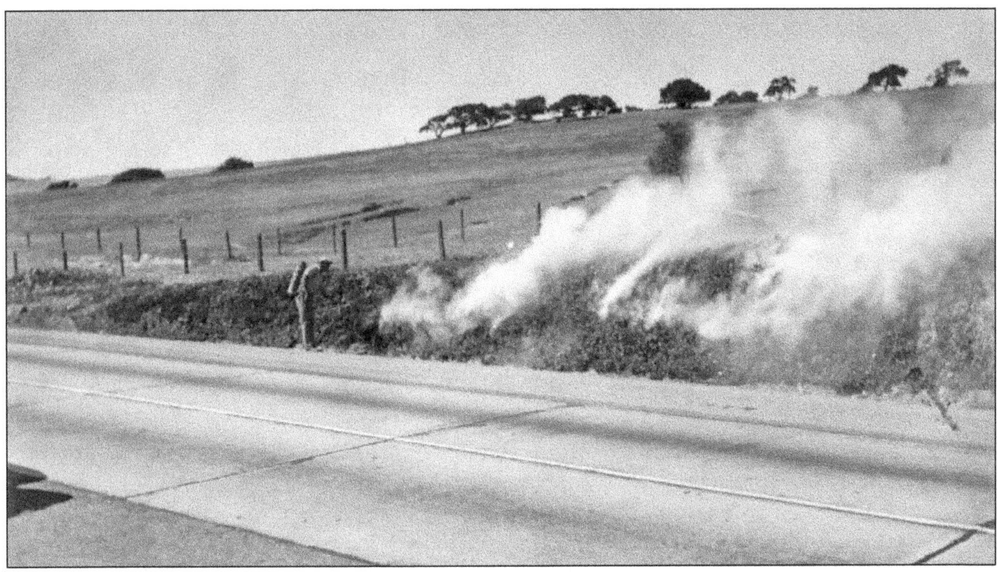

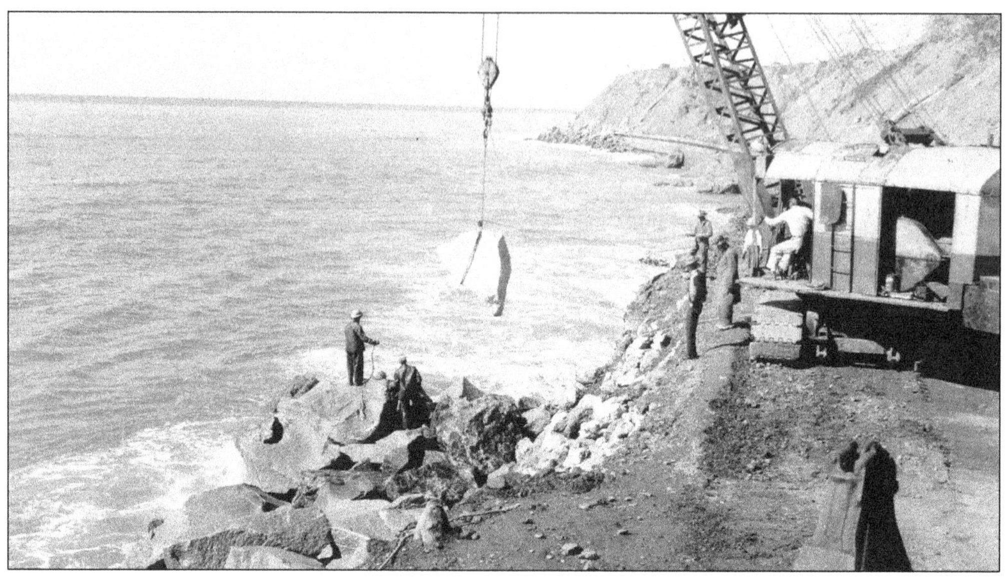

Road construction workers use a crane to lower large boulders into the ocean as shore protection along a section of the Ventura coast by Point Mugu in 1945. Shore protection helps defend the highway from tide and wave action. The drawback is that wave reflection from man-made shore protections can lower the sand level of the fronting beach and cause erosion to nearby unprotected coastal areas that lack shore protection. A significant amount of boulders and rocks used for shore protection along the Ventura coast came from the portion of the Santa Monica mountains that was blasted away to create the Point Mugu passage.

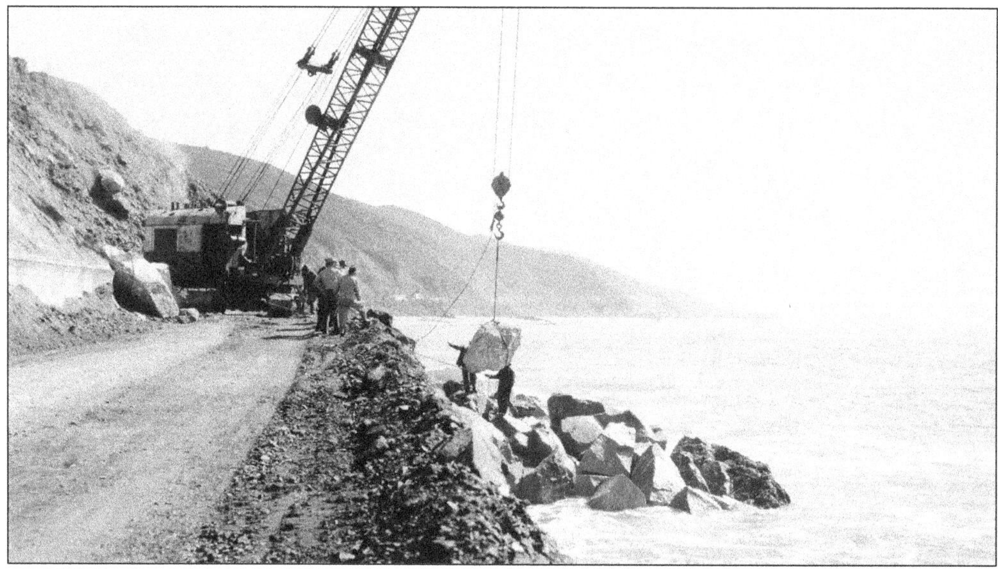

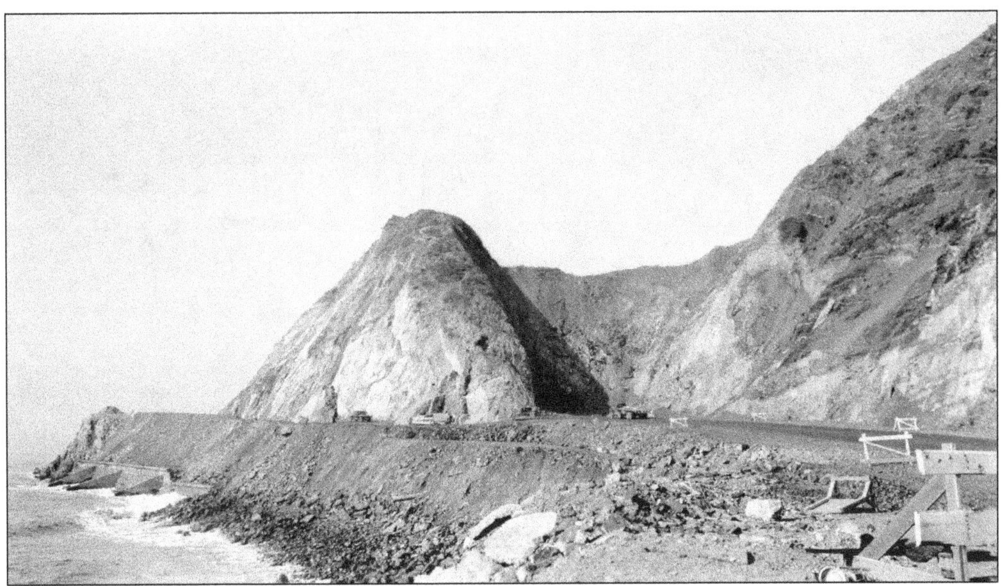

A portion of the Santa Monica Mountains at Point Mugu is seen in this 1945 image before a passage was cut out. A promontory of solid rock, the area had to be surveyed from motorboats. The only means of bringing in heavy equipment was on ocean barges, which were then floated up to the site. Massive amounts of solid rock and dirt had to be dug out and blasted. Up to 107 tons of explosives were used to create a passage.

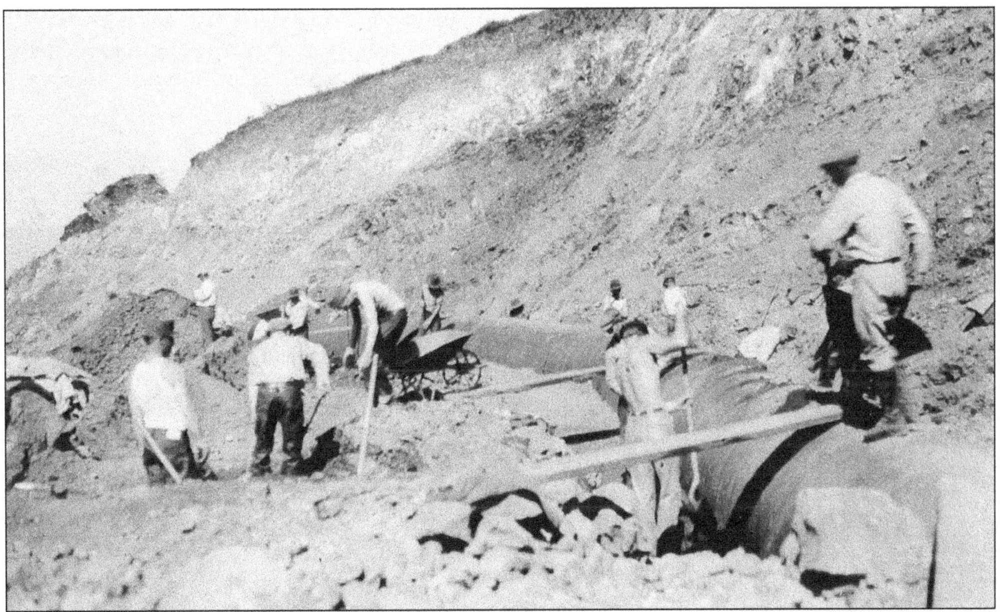

This 1932 photograph shows laborers using hand shovels to backfill an installed culvert at Rat Canyon Creek in Monterey. The culvert will allow water to flow under the highway from one side to the other, which is necessary at natural drainage and stream crossings. Culvert failures occur when a culvert is too small to handle a flood event or when the flood compromises the road above the culvert.

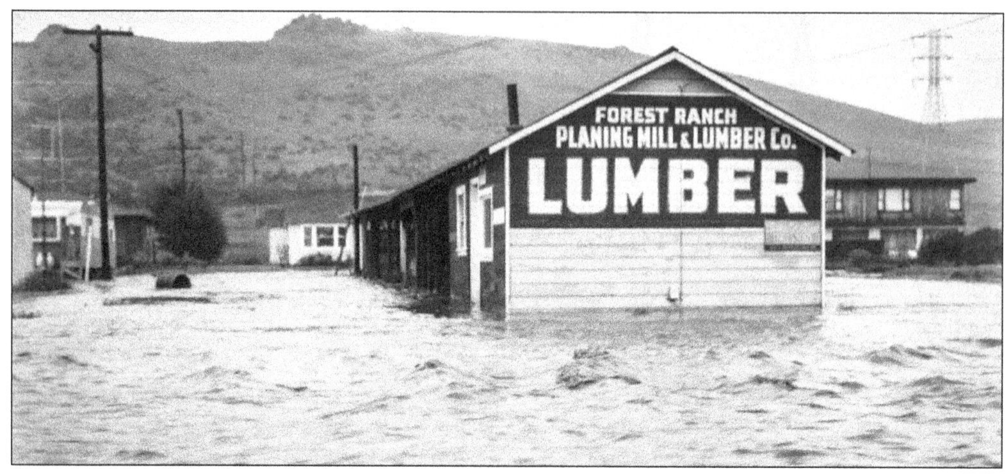

In addition to landslides, flooding also damages roads and highways. In 1956, a strong storm with exceptional rainfall flooded the town of Morro Bay, causing the Morro Creek to overflow. Flooding can wash out roads and damage culverts, and high, raging water can destroy structures. A heavy downpour along the coast can produce flash floods causing mudslides that blanket roads. Below is an upstream side view of Morro Bay Creek Bridge at peak flow during the 1956 flooding. Flooding can also compromise bridges that are susceptible to scouring, which is the erosion of the stream bed material caused by flowing water. It can wash away large sections of foundation material from under a bridge's footings, resulting in the bridge becoming unstable.

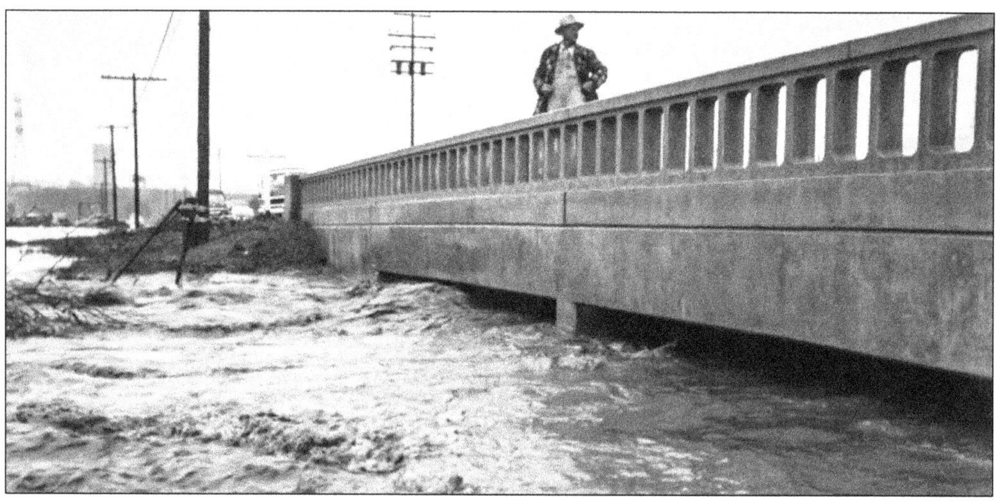

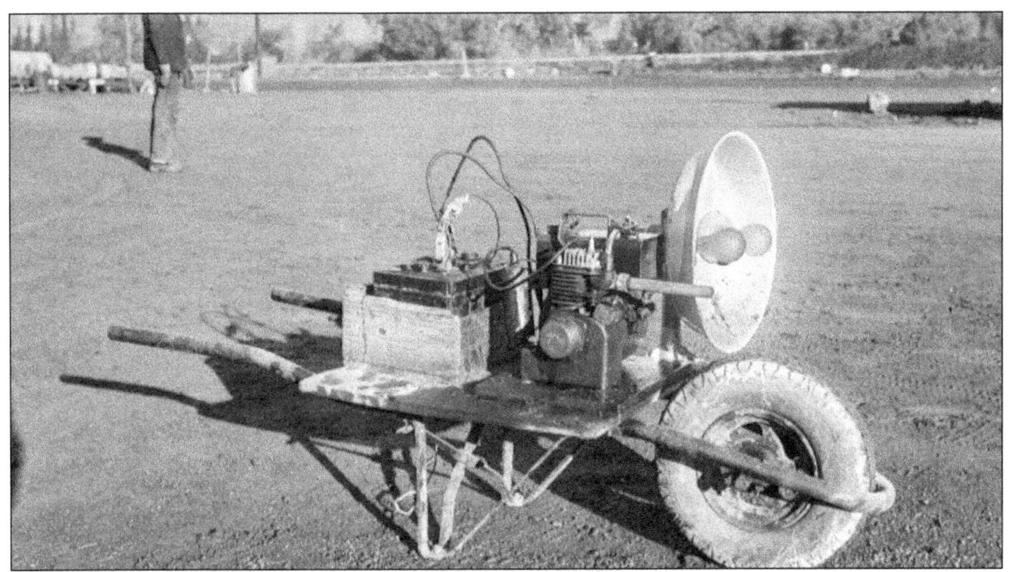

The portable Baby Delco Light Plant, seen above, was used to provide light where needed during road and tunnel construction. There were approximately 150 companies manufacturing light plants, but the Delco light was one of the most popular and successful systems. Delco manufactured light plants for the war effort during World War II but ceased production of the Delco Light Plant by 1947, when electricity was made available in most rural areas. Below is a 1939 double cylinder or compound three-wheel steam roller. It was used for base compaction through bounce and vibration.

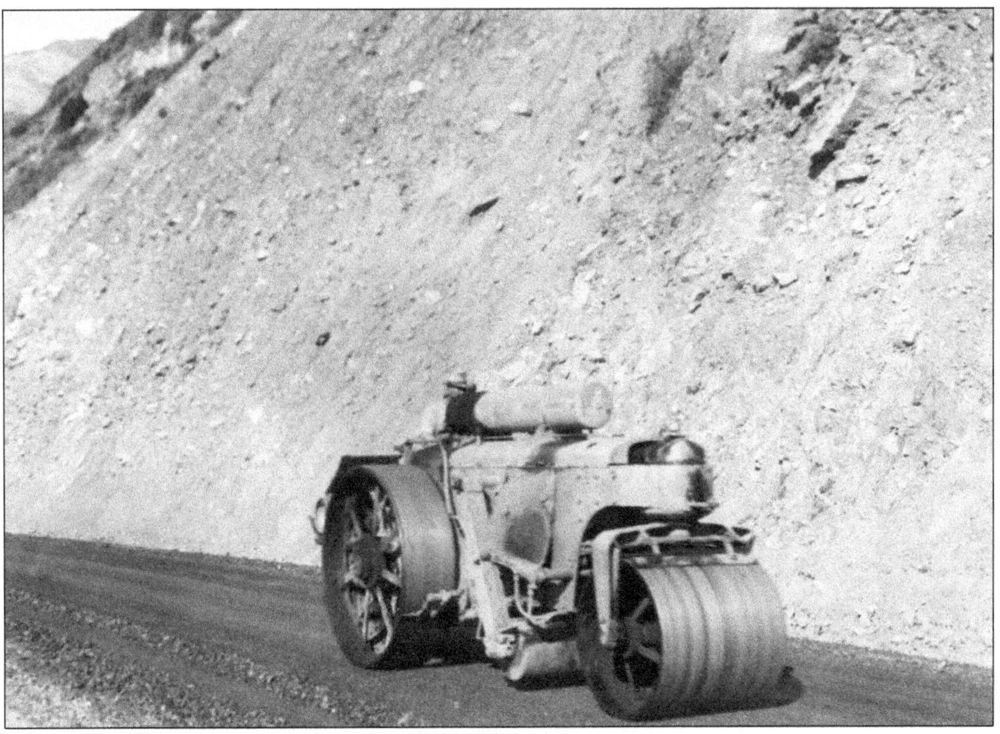

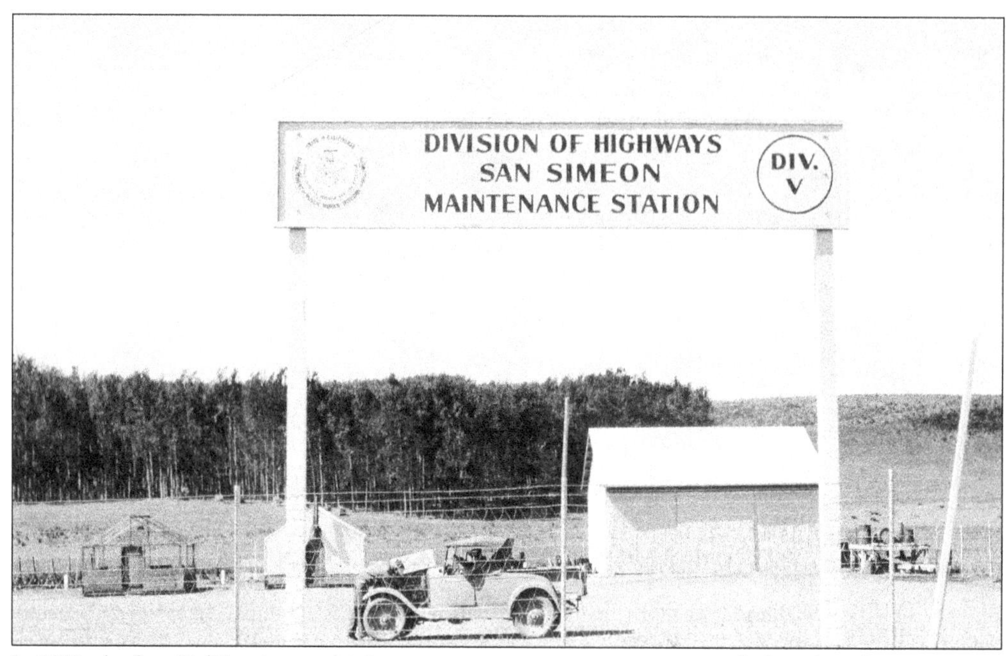

In 1912, the first California state highway construction contract was awarded, and road construction commenced the same year. The contract award included the construction of coast roads, and the Department of Engineering was responsible for road building projects. It was later renamed the Division of Highways and placed under the Department of Public Works. In 1923, the Department of Public Works underwent a reorganization, and the Division of Highways became the responsibility of the California Highway Commission. The reorganization created six new divisions, including the San Simeon Maintenance Station Division V. Division V comprises the counties of Santa Barbara, San Luis Obispo, Monterey, San Benito, and Santa Cruz, and is responsible for all road construction, maintenance, and projects in its jurisdiction. Below is a 1933 photograph of employee K. Donovan sitting at her desk in the San Luis Obispo office.

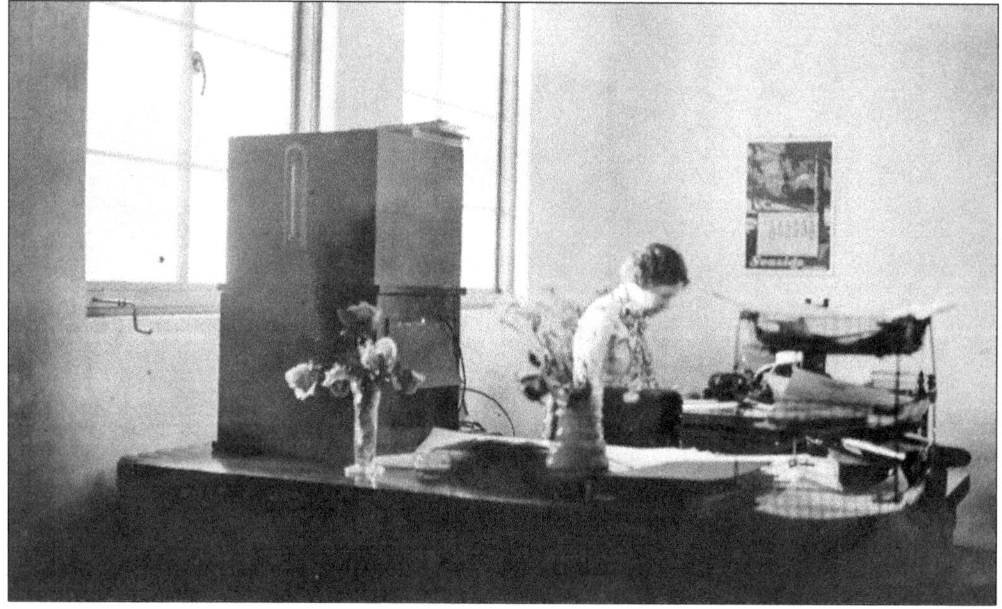

Two

Coast Bridges, Tunnels, and Lighthouses

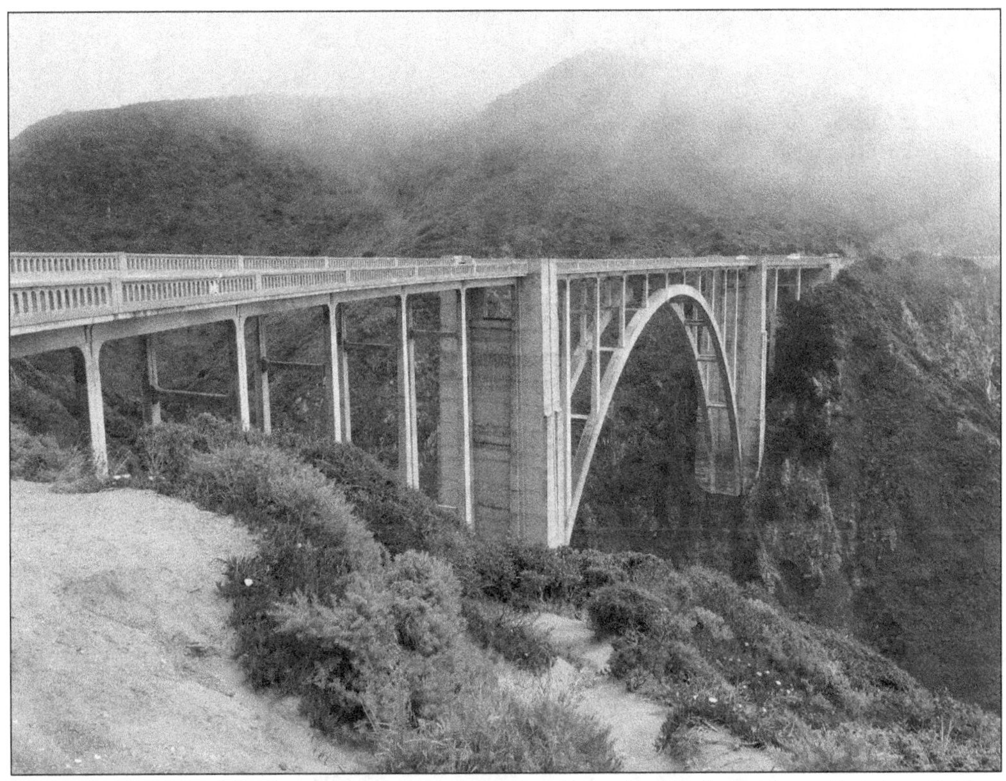

Pictured is the famed Highway 1 Bixby Creek Bridge, also called Bixby Bridge or Rainbow Bridge, located on the Big Sur Coast. Its construction was an engineering marvel of its time, and it remains in operation after 85 years. It is one of the most photographed bridges in the world and has been featured in many movies and commercials. Prior to its construction, the Big Sur area was isolated from coastal communities to the south. (Robert C. Pavlik.)

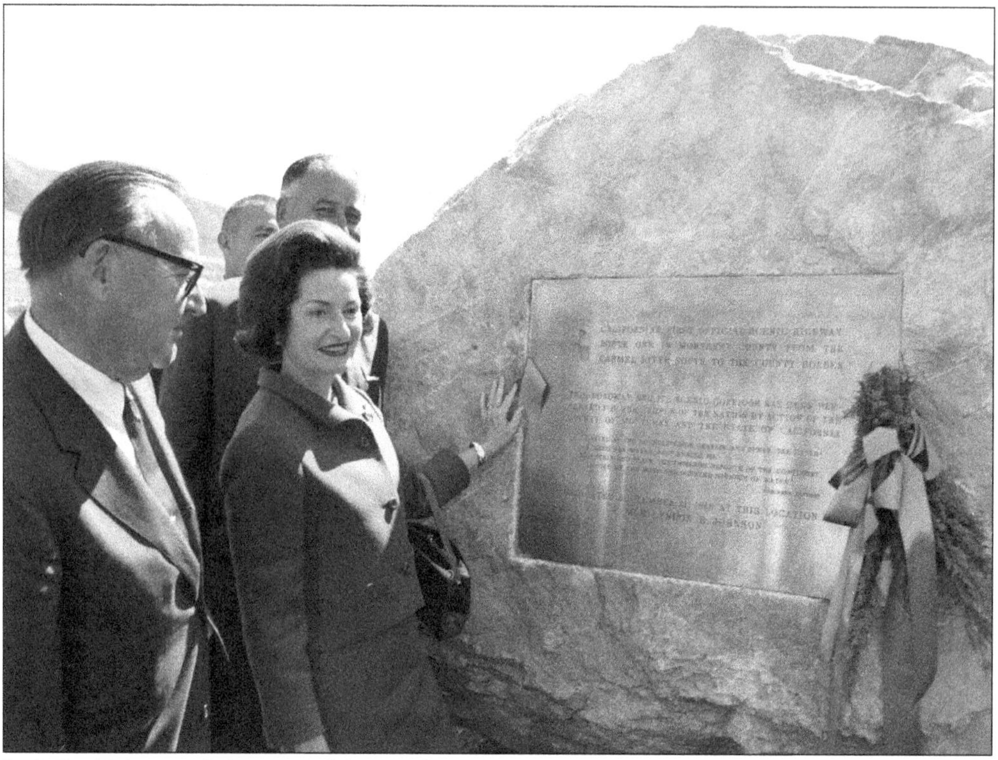

Lady Bird Johnson dedicates California's first official scenic highway atop the Bixby Creek Bridge Scenic Overlook in 1966. Standing behind her is California governor Edmund "Pat" Brown (to her left) and former California state senator Fred Farr. The Highway Beautification Act was signed in 1965 by Pres. Lyndon Baines Johnson, who stated, "Beauty belongs to all the people." The act was passed to preserve the natural beauty of America and the lands adjoining its burgeoning highway system. Seen below is a replica plaque that was created in 2015 by artist Gustavo Torres to replace the original, which went missing. A 50-year rededication ceremony was held in 2015. Luci Baines Johnson, daughter of Lady Bird and Pres. Lyndon Baines Johnson, and Sam Farr, son of Sen. Fred Farr, attended the event. The plaque was donated by the LBJ Foundation and is mounted next to the Bixby Creek Bridge. (Below, Robert C. Pavlik.)

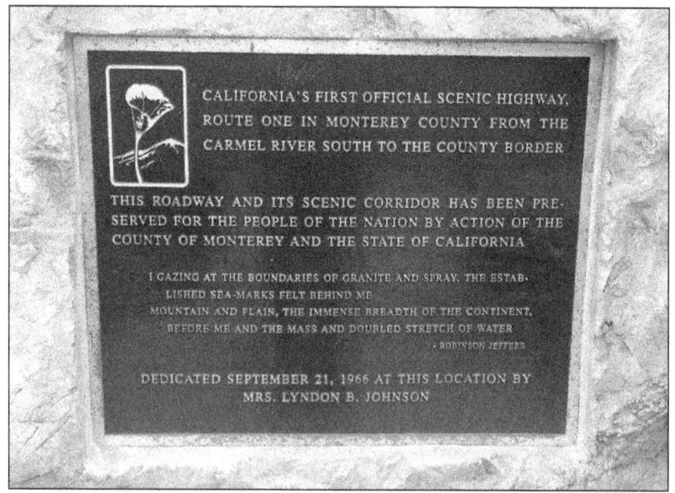

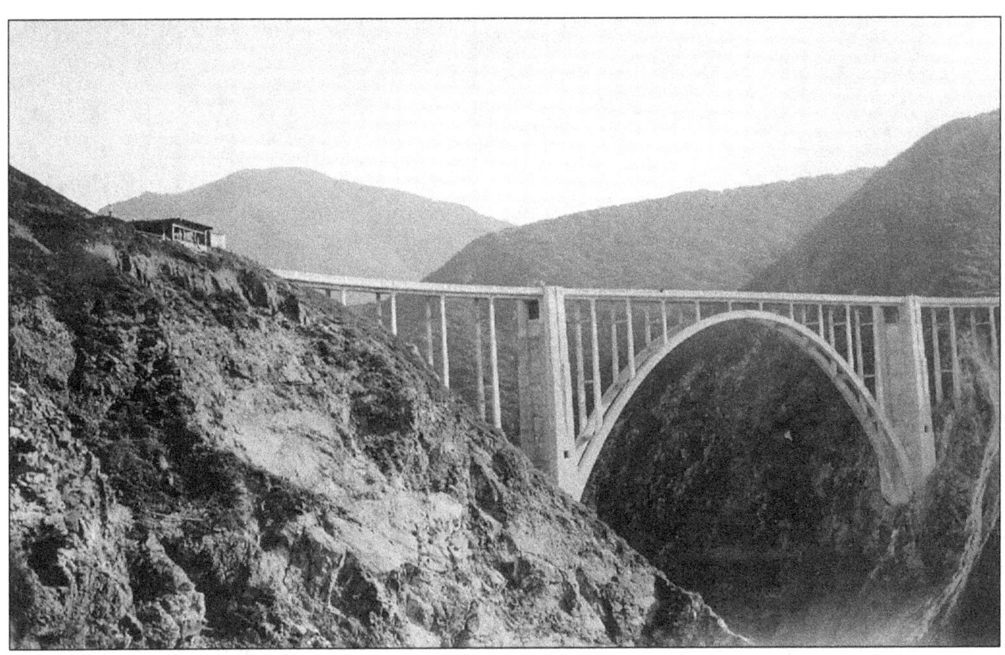

The Bixby Bridge was constructed on a curve above a canyon between majestic mountains and the sea. Its location was specifically chosen for its sensational and dramatic backdrop. Bridge designers agreed that an arch design would go well with the configuration of the canyon, and concrete would blend nicely with the landscape. Standing at 320 feet, the central arch span was once the longest in California. Completed in 1932 at a cost of only $200,000, it still stands strong and is known worldwide for its remarkable location along the Pacific Coast Highway. Below is another angle of the bridge showcasing the remarkable design and construction that enabled the San Simeon to Carmel section of Route 1 to become California's first scenic highway.

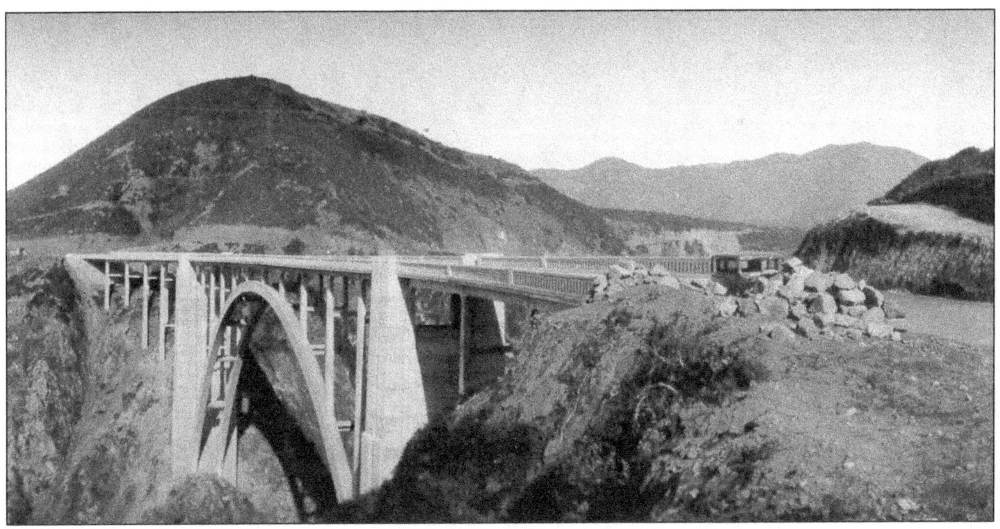

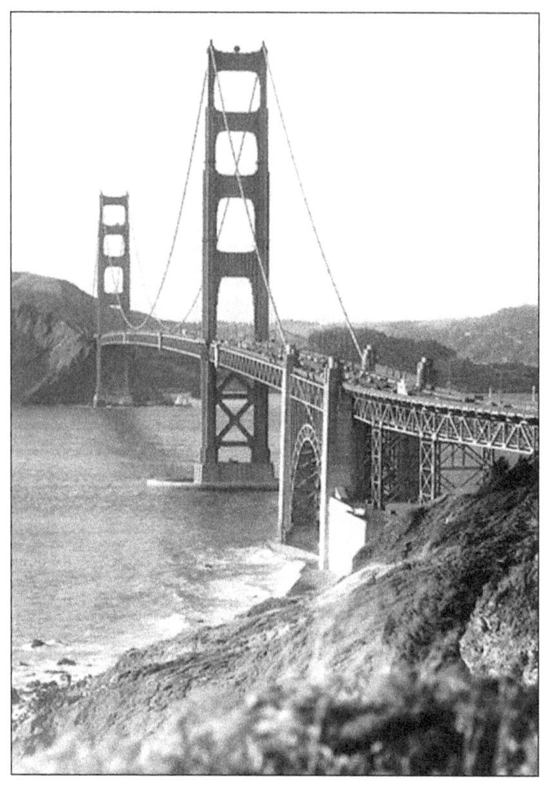

San Francisco's Golden Gate Bridge is a simple suspension bridge with a main span of 4,200 feet; at the time of its completion, it was the world's longest bridge. Its length from abutment to abutment is 1.7 miles. The original combined weight, including anchorages and approaches, was 894,500 tons. Its two main towers support the two main cables. The height of the tower above the water is 746 feet. The two main cables pass over the towers and are secured at each end with giant concrete and steel anchorages. The length of one main cable is 7,650 feet, with a diameter of 36.375 inches. Construction began in 1932, and the bridge was completed and opened in 1937. Below, hundreds of people celebrate the opening of the Golden Gate Bridge on May 27, 1937. (Left, C. Hanchey.)

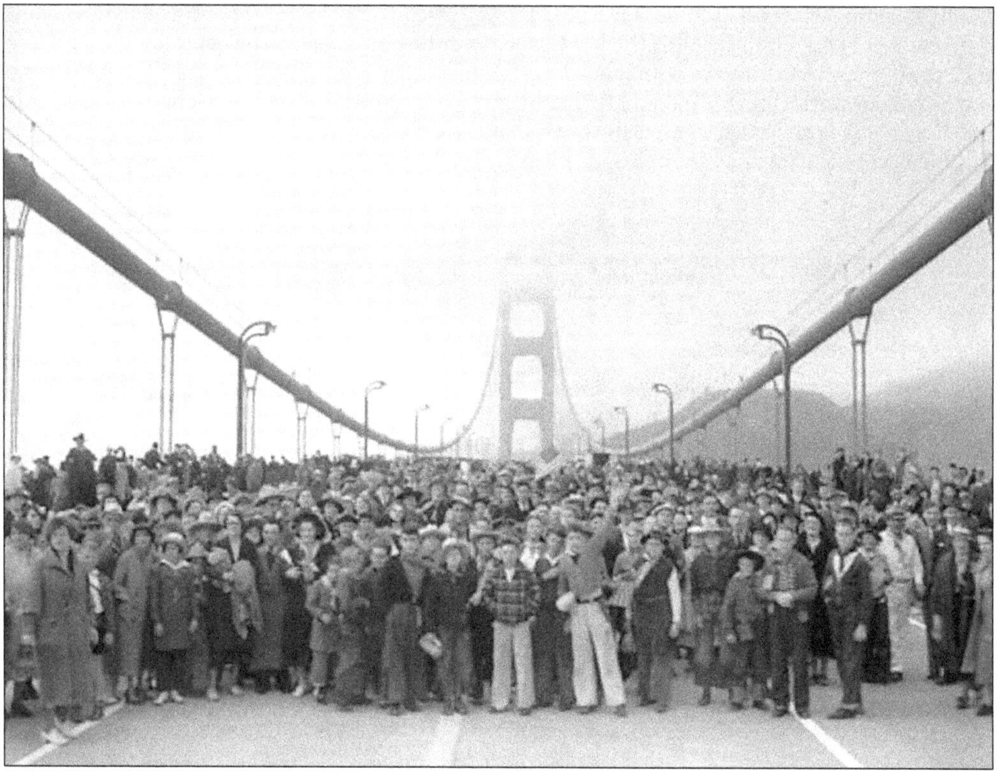

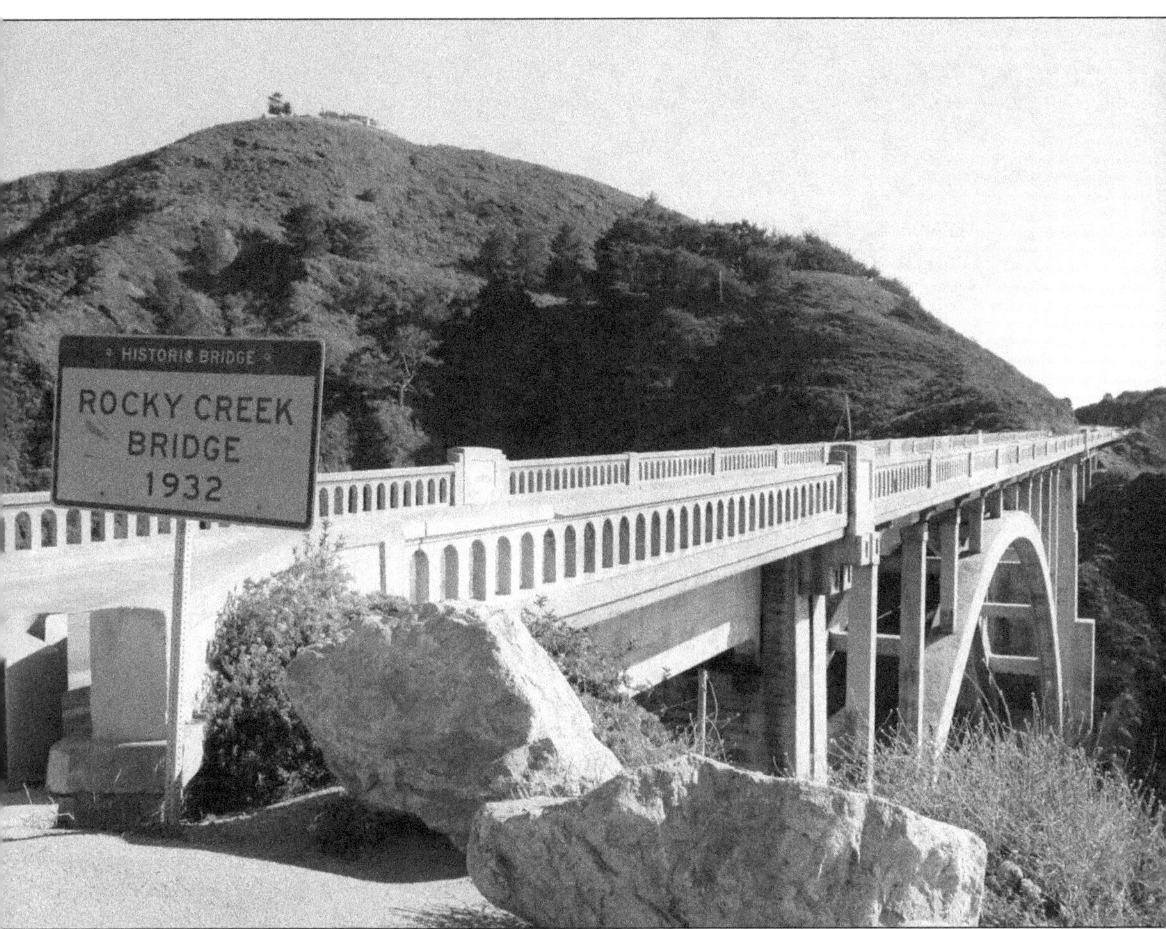

The Rocky Creek Bridge historic bridge sign is shown here. Rocky Creek Bridge looks similar to—and is sometimes mistaken for—the Bixby Bridge because of its open-spandrel arch. It was one of several bridges constructed in the 1930s during the road building project along the Monterey coast. It is also one of six bridges along the Big Sur Coast that were constructed above deep canyons. It sits high above the Rocky Creek stream canyon at 500 feet. It was constructed of reinforced concrete with a 239-foot central span; the bridge itself is 150 feet. The curve at the end of the bridge shows the engineering genius it took to construct a bridge above deep canyons and winding roads. (C. Hanchey.)

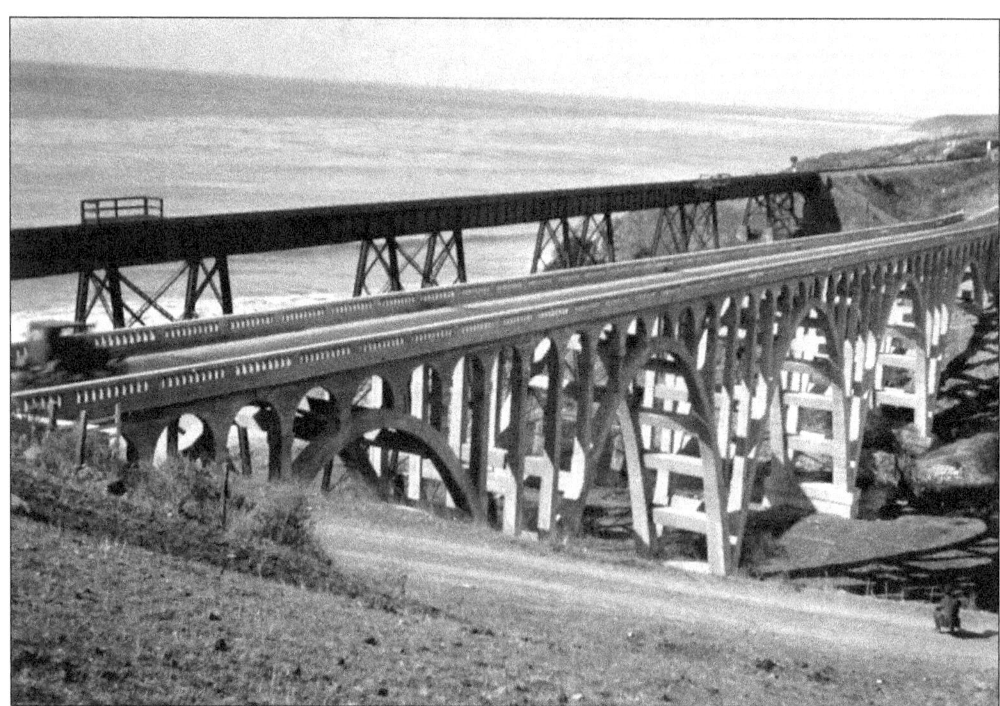

Arroyo Hondo Bridge in Santa Barbara was one of the first complex bridges constructed of concrete and steel by the California Division of Highways in 1918. Standing at 529 feet, it was open for more than half a century until the highway was bypassed in 1984. The old Southern Pacific trestle ran parallel to the bridge.

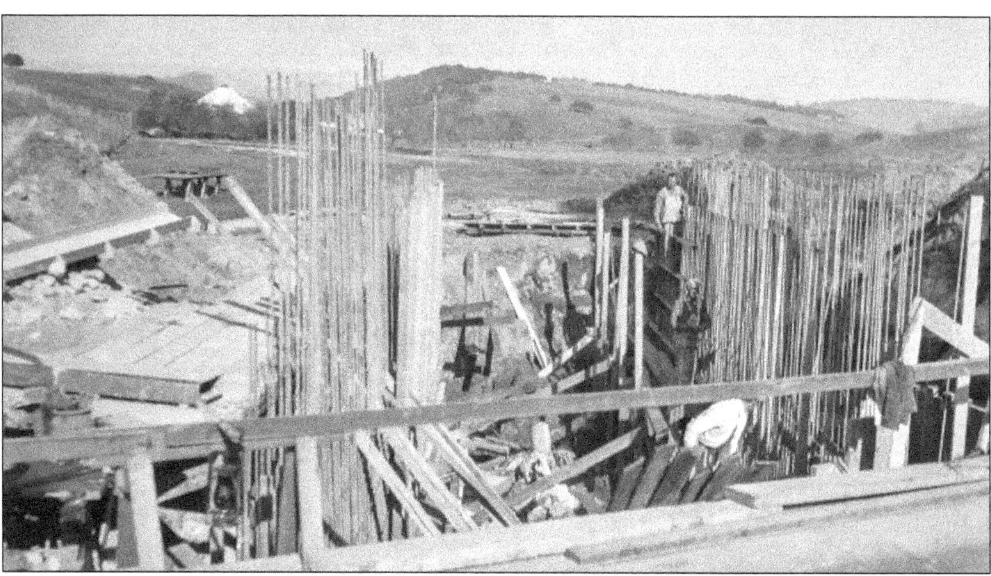

These bridge workers in the 1930s are constructing falsework, which is a temporary structure to support arched or spanning structures for formwork to mold concrete to a desired shape. A scaffold can be seen, which gives the workers access to the structure being constructed. Timber was used in early falsework and scaffolding.

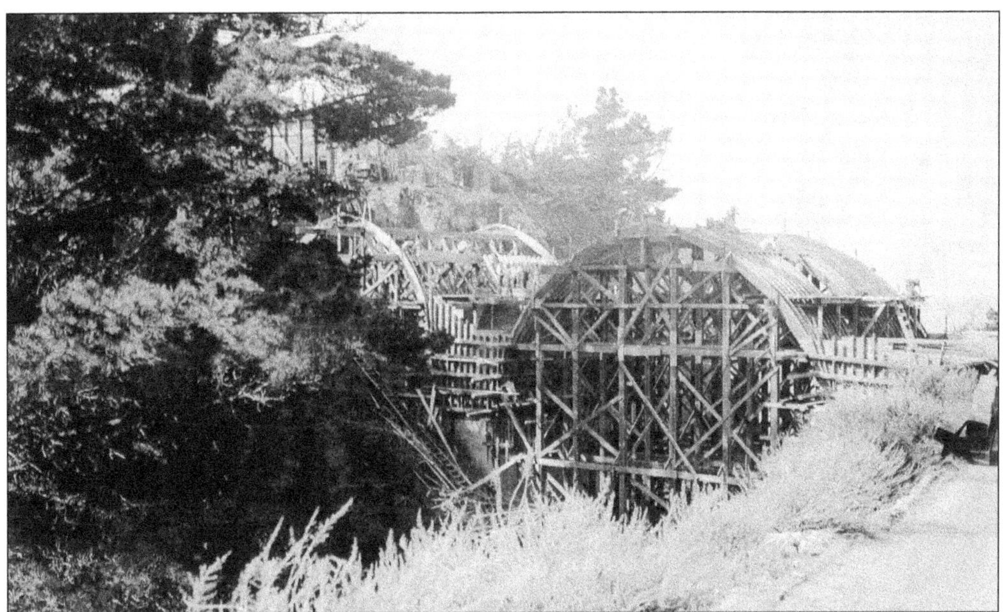

A completed formwork and falsework are visible in this 1933 photograph of the construction of the arch at Wild Cat Canyon Bridge in Monterey. The shape of the bridge is created by the formwork; the falsework provides the support needed to enable the liquid concrete to set and cure.

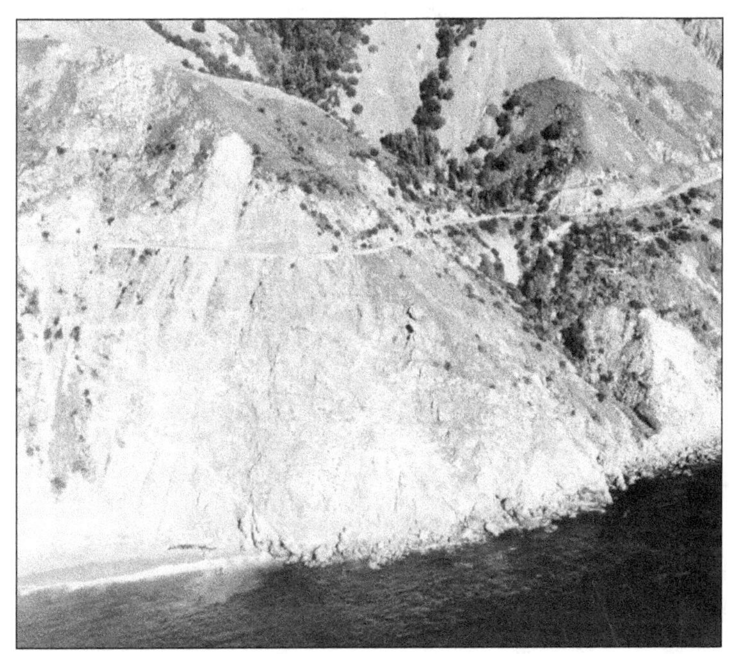

This is a 1963 aerial view of a two-lane segment of Route 1 alongside a section of the Monterey coast mountains atop steep cliffs with the ocean below. The Big Sur Coast area was virtually isolated until the Carmel–San Simeon Highway was completed in 1937. (Copyright 1963 California Department of Transportation. All rights reserved.)

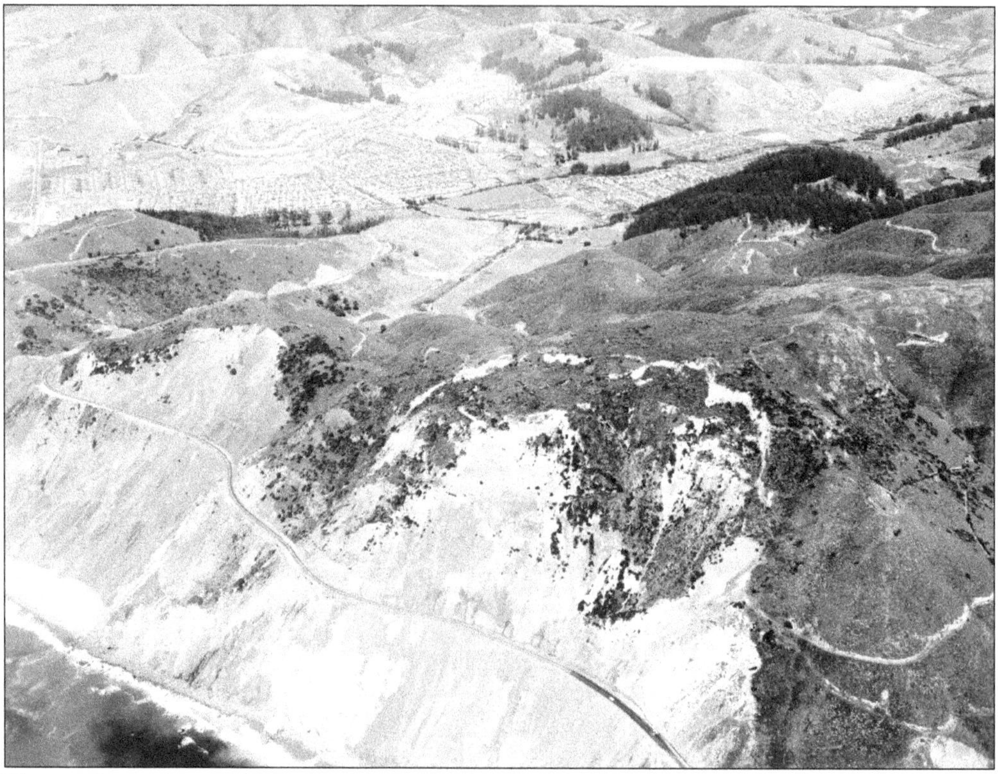

This 1959 aerial view shows another angle of a two-lane stretch of Route 1 winding atop steep cliffs above the ocean. Although ominous, the highway along Monterey's coast, particularly the Big Sur Coast, is regarded as the most picturesque stretch of Route 1. (Copyright 1959 California Department of Transportation. All rights reserved.)

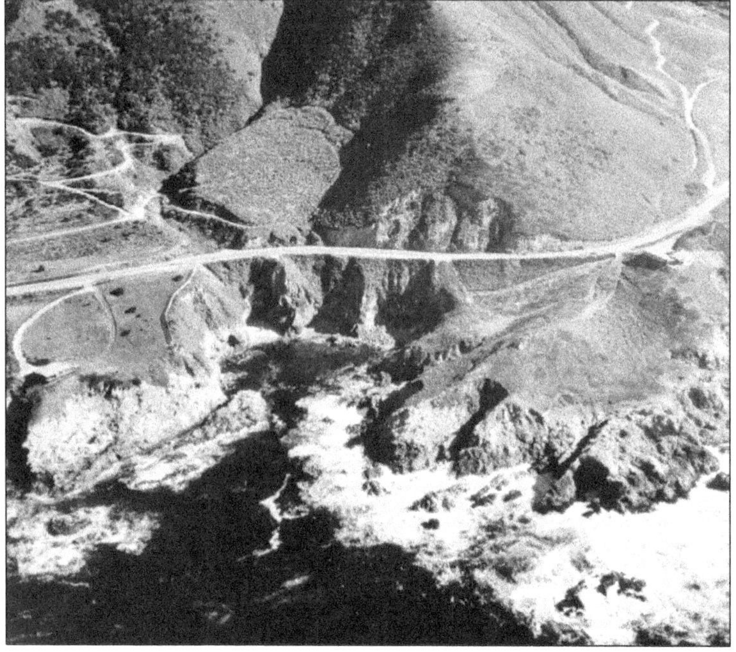

This aerial view of the Big Sur Coast in 1963 showcases its dramatic mountains and rugged rocky promontories, caves, and tide pools. The hundred-mile stretch of coastline between Carmel and San Simeon is commonly referred to as the Big Sur Coast. (Copyright 1963 California Department of Transportation. All rights reserved.)

In this 1960 image, the Pacific Coast Highway runs along the Ventura County coast. Sandwiched between rough mountainsides and the sea, sand and dirt can be seen spilling onto the highway from both sides along a stretch where there are no protective shoulders. (Copyright 1960 California Department of Transportation. All rights reserved.)

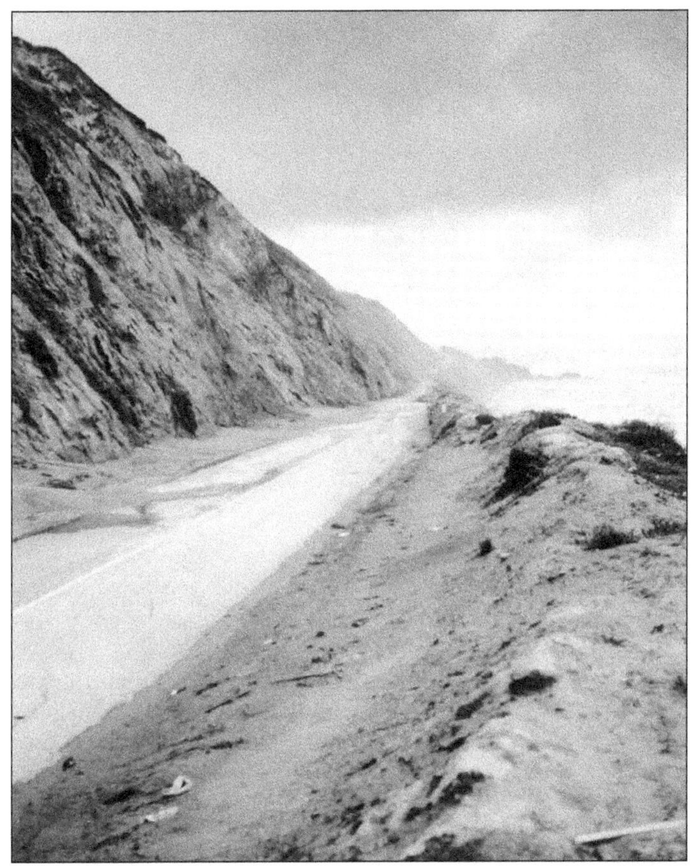

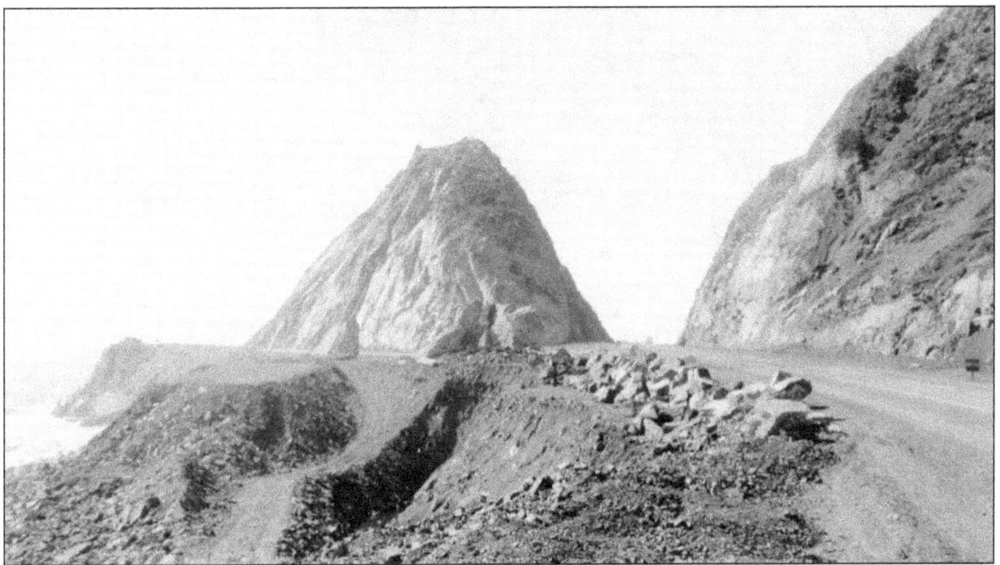

This 1940 photograph shows the passage blasted through the rocky promontory of Point Mugu. The blasting left one side of the mountain of solid rock at the shore's edge intact, which was given the name Mugu Rock. (Copyright 1960 California Department of Transportation. All rights reserved.)

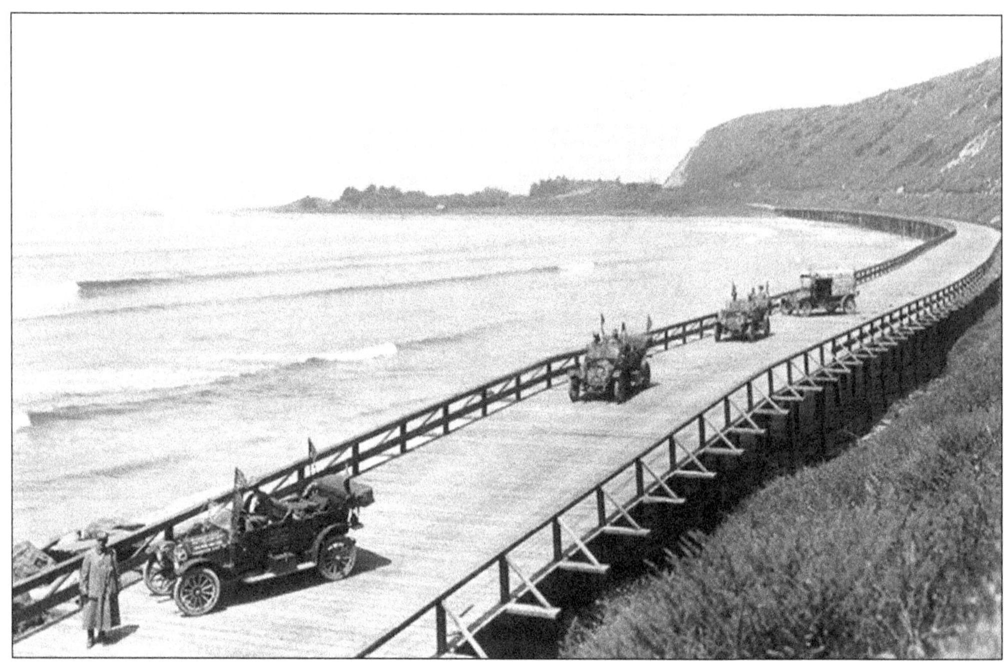

The elevated Rincon Causeway in Ventura County is seen here in 1913. Also called Rincon Sea Level Road, it was part of the coast route between San Francisco and Los Angeles. It is now part of Route 1 and is also called the Rincon Parkway. It was constructed entirely of wood. Below is another angle of the Rincon Causeway looking south. It was the first coast automobile route that allowed motorists to traverse large sections of the California coast from San Francisco to Los Angeles. Continued maintenance of the causeway was necessary due to waves hitting the pilings, especially during strong storms.

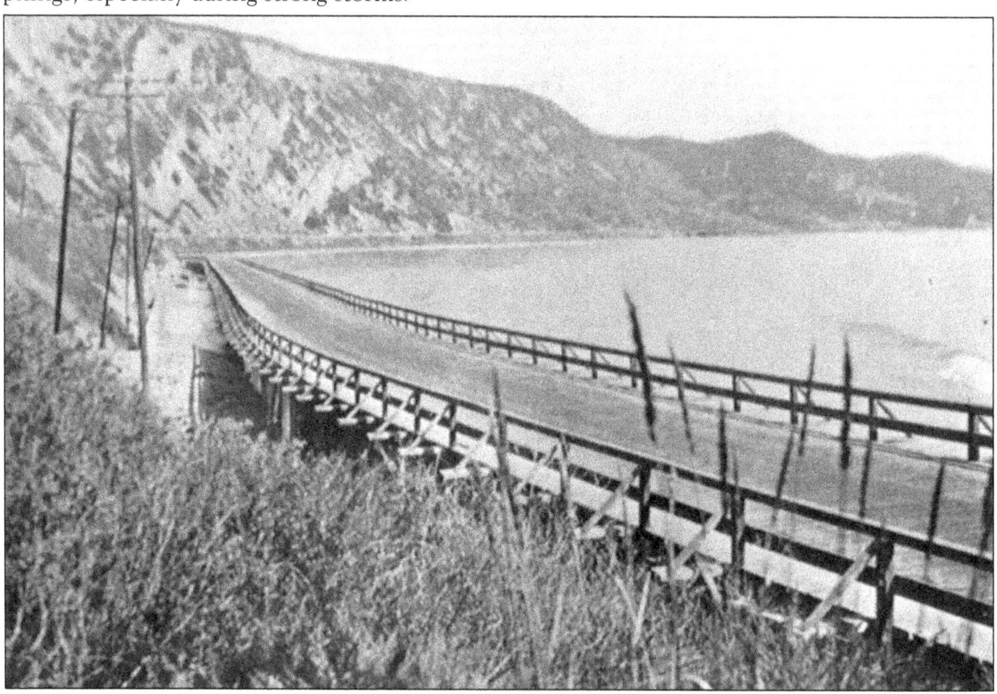

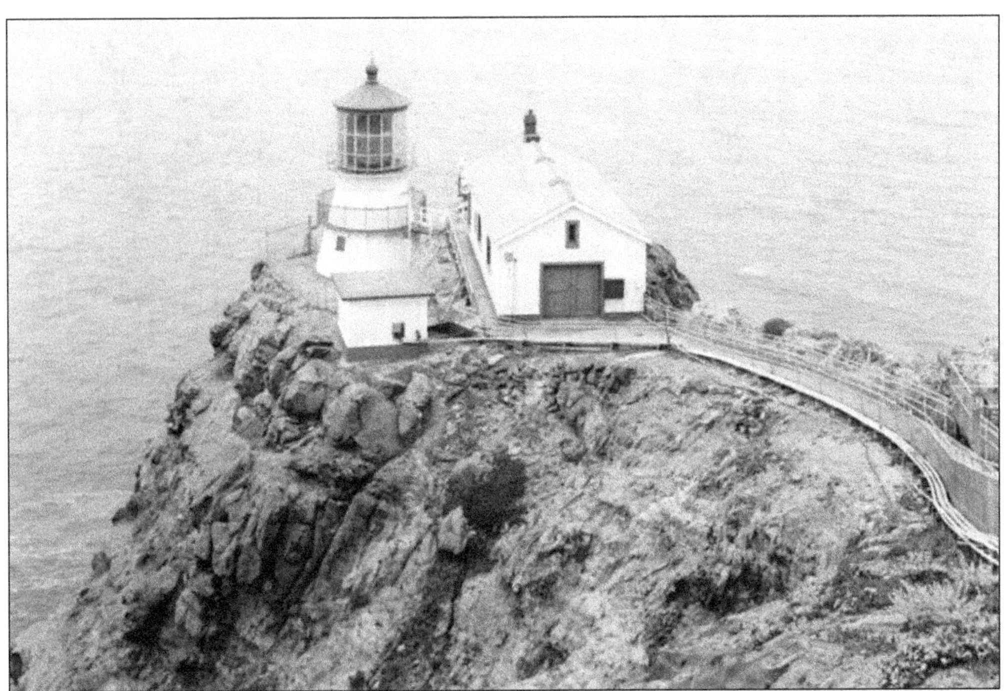

Point Reyes Lighthouse in Marin County is pictured here in 1933. Before modern electronic navigational systems were invented, lighthouses were used as navigational aids for mariners. They marked dangerous coastlines, warned of shoals and reefs, and showed safe entries to harbors. Built in 1870, Point Reyes Lighthouse was retired from service in 1975. (Library of Congress.)

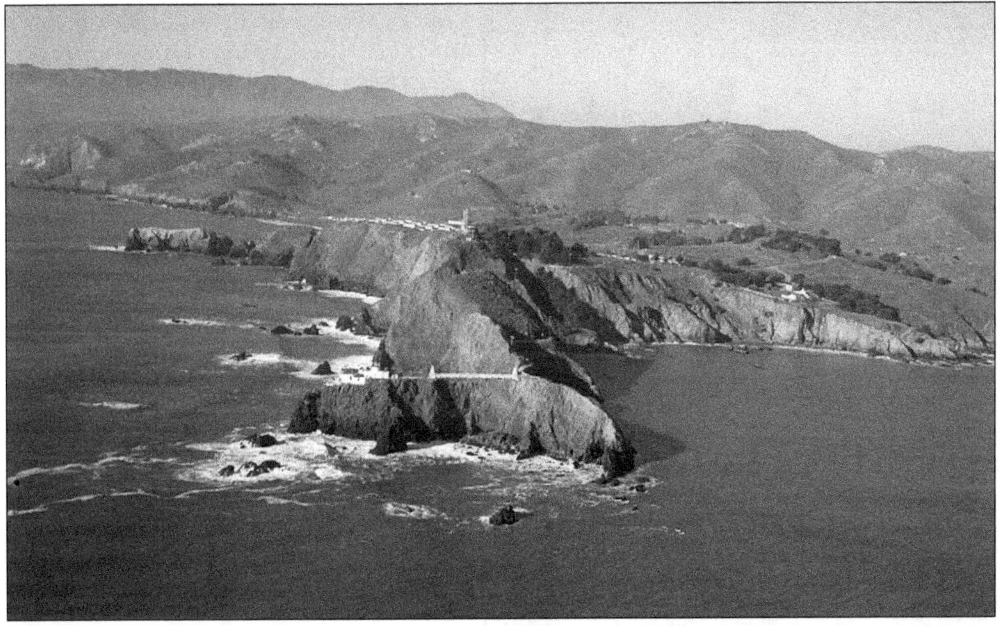

Point Bonita Lighthouse at the entrance to San Francisco Bay is seen here. Built in 1885, the original lighthouse was 306 feet above sea level. Due to the prevailing clouds and fog, it was too high to be of use to mariners. This is its second location at 124 feet. It was the last manned lighthouse on California's coast. (Library of Congress.)

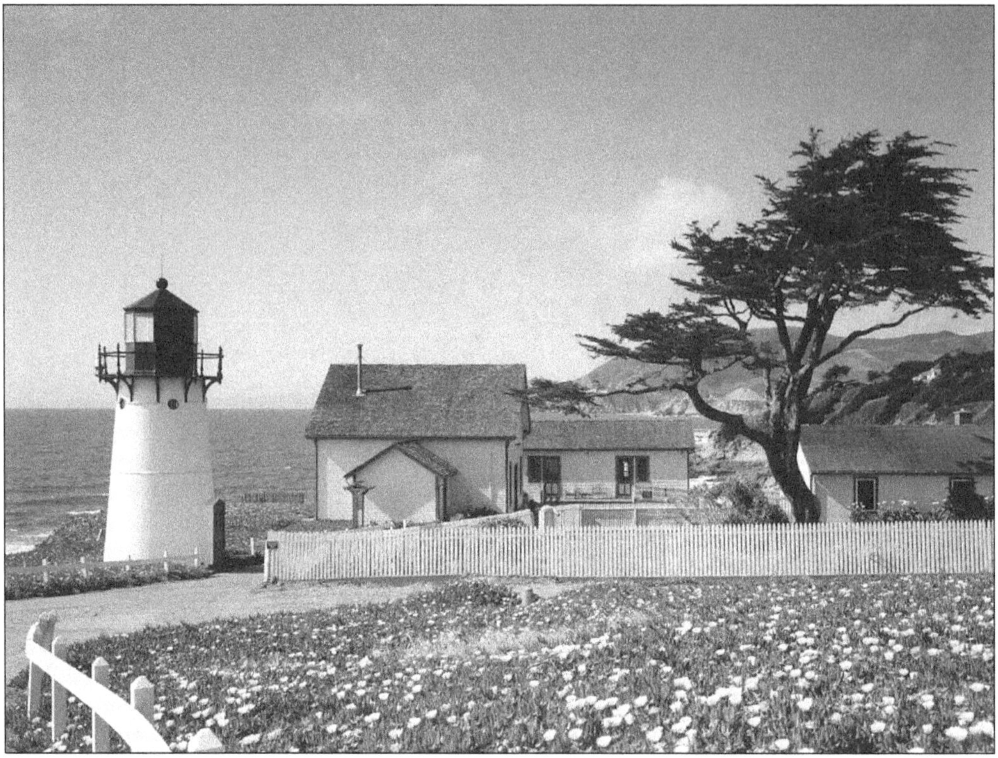

Point Montara Lighthouse in San Mateo County is seen here in 1980. It was originally built in Massachusetts in 1881 and was shipped to California in 1928 to replace the original lighthouse. It stands 30 feet tall and is still in service. (Library of Congress.)

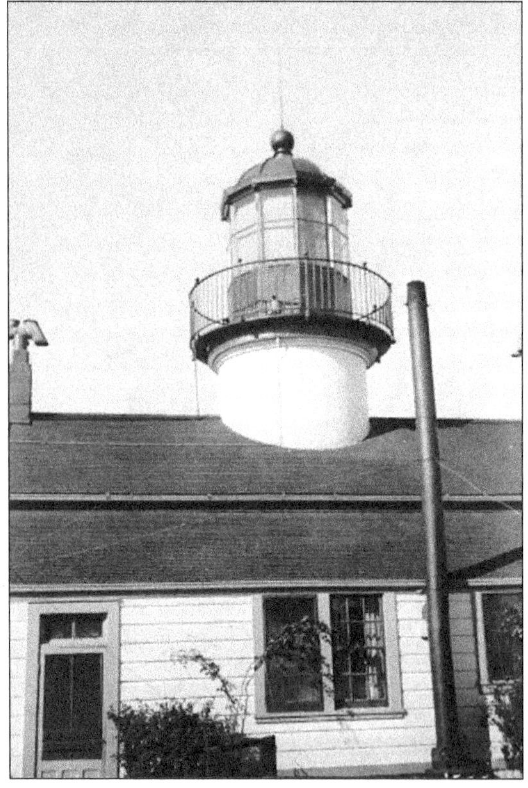

Point Pinos Lighthouse was built in 1855 and remains in service. Located on the coast on the Monterey Peninsula, it is the oldest lighthouse operating on California's coast. Its light tower is 89 feet above sea level. Rising out of a bungalow structure, its style differs from most lighthouses. (Library of Congress.)

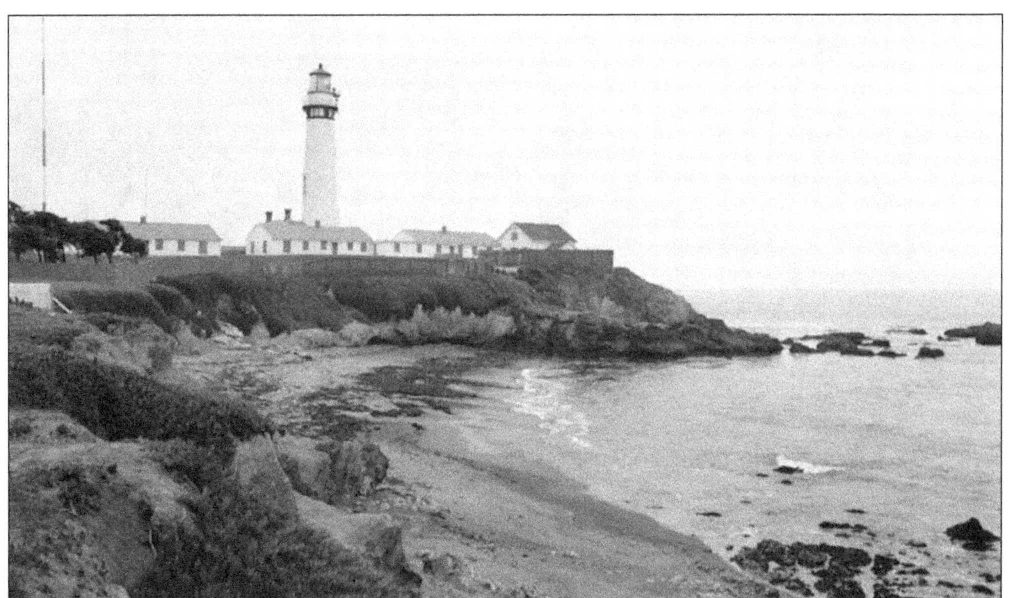

Pigeon Point Lighthouse in San Mateo County has helped mariners navigate since 1872 and is still in service. It is widely known for being one of the tallest lighthouses in the country. It has a Fresnel lens with 1,008 prisms, and its five-wick whale oil lamp was first lit on November 15, 1872. (Library of Congress.)

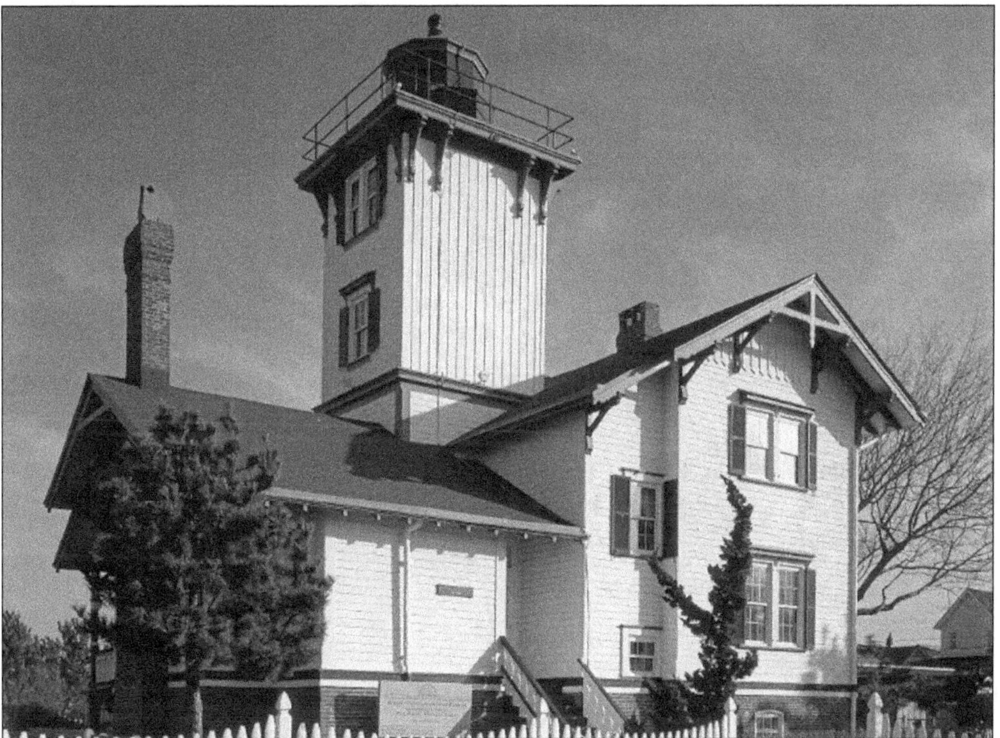

In Southern California, Point Fermin Lighthouse in San Pedro Bay was designed by Paul J. Pelz and was built in 1874. It was the first light in San Pedro Bay, which later became the main port in Los Angeles County. (Library of Congress.)

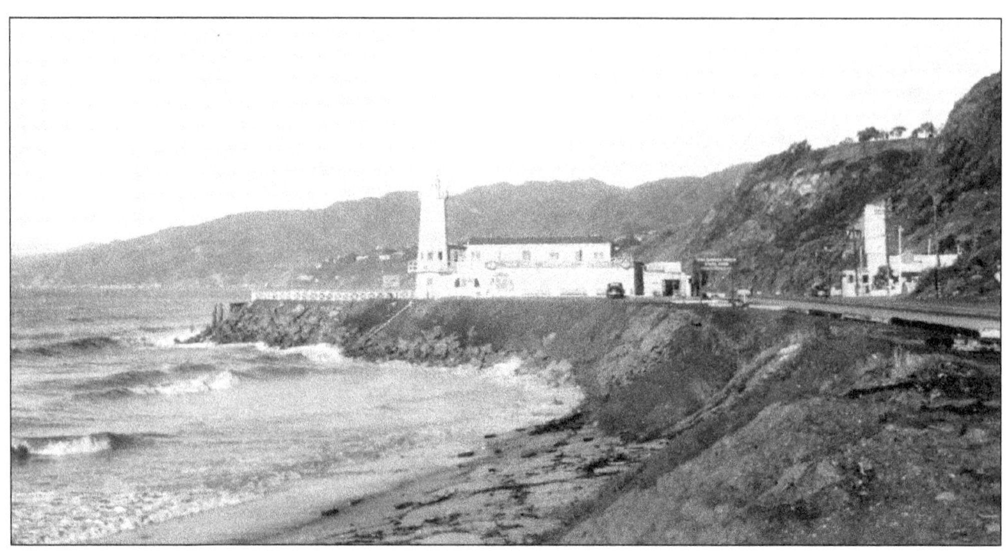

The Santa Monica coast lighthouse was built in the early 1900s and later marked the site of Long Wharf at Portero Canyon. Plans to save the lighthouse by relocating it to Pepperdine University off the Pacific Coast Highway in Malibu failed when the aged tower was unable to survive the move.

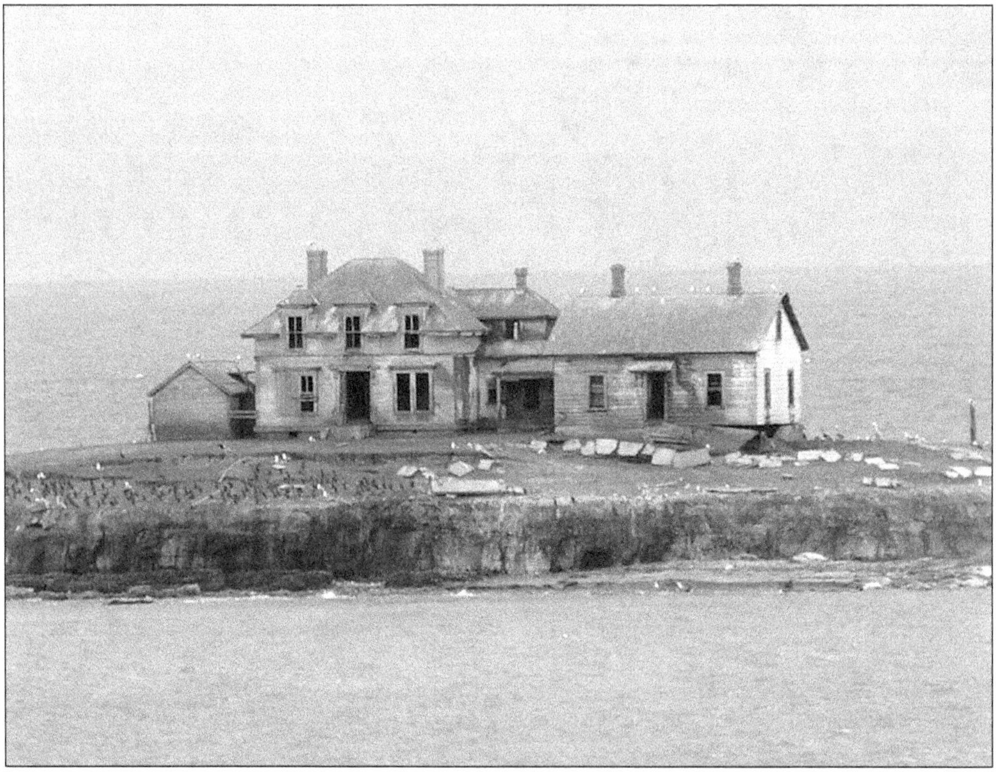

Located on Ano Nuevo Island, Ano Nuevo Lighthouse was in service from 1872 to 1948. It once warned mariners away from the treacherous low rocks along nearby Points. It is now abandoned. Shipwrecks necessitated the lighthouse, notably the *Carrier Pigeon* in 1953, and *Sir John Franklin* in 1865. (Jef Poskanzer.)

Established in 1889, Point Sur Lighthouse is perched on a 361-foot rock. Shipwrecks occurred at the point in the mid-19th century, notably the *Ventura* in 1875. Before completion of Highway 1, the area was virtually isolated. Facilities were built at the station to sustain lighthouse keepers and their families. (US Coast Guard.)

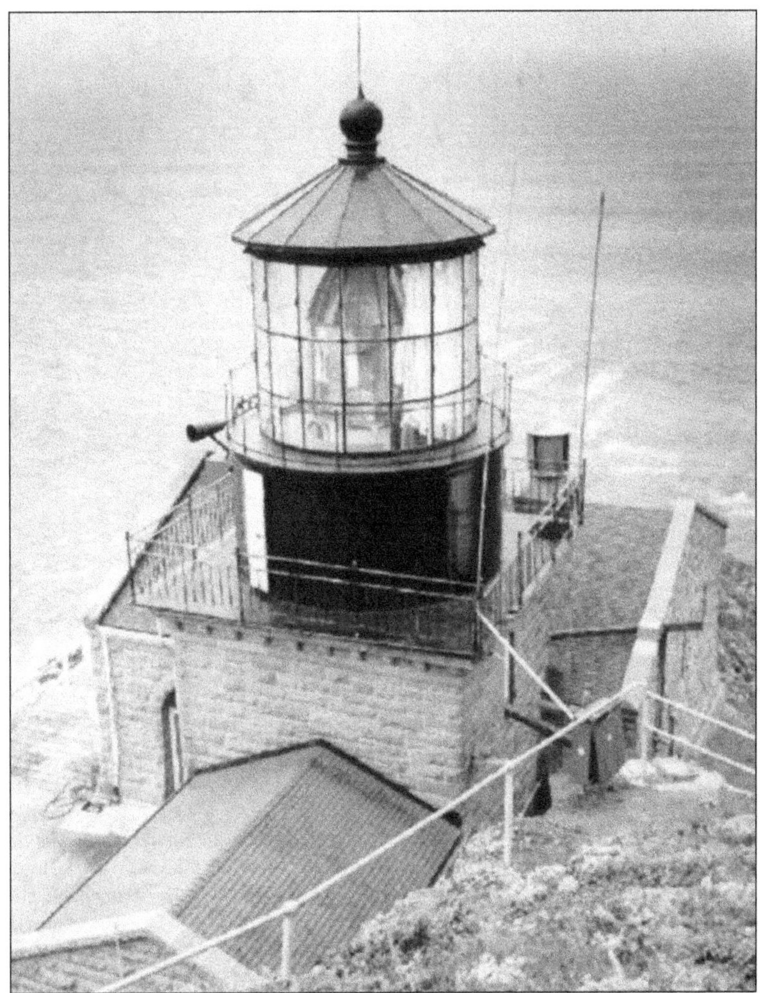

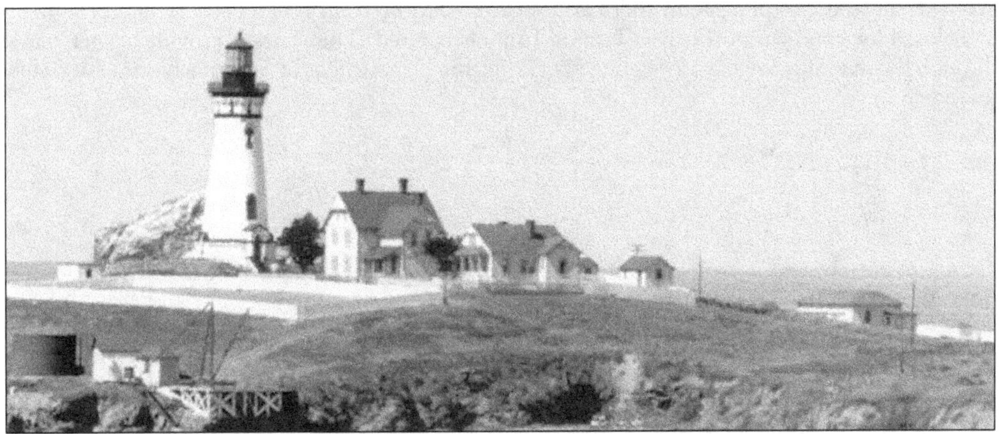

On February 15, 1875, Piedras Blancas's first-order Fresnel lens was illuminated and is still in service. Originally standing at 100 feet, earthquakes necessitated the removal of the upper three floors, reducing it to 70 feet. It was once one of the three tallest lighthouses in California. (Hearst Castle.)

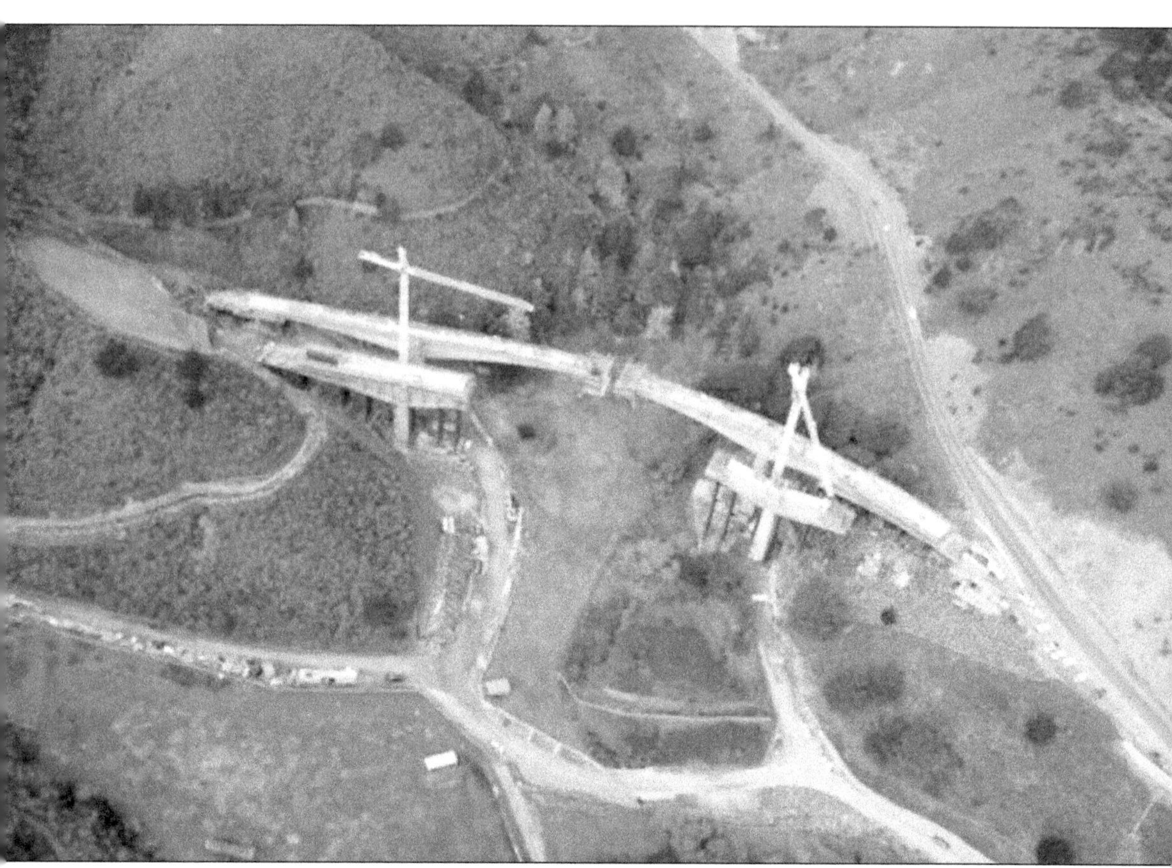

This is an aerial photograph of the construction of two side-by-side curved bridges that will connect Highway 1 to the north portal of the Tom Lantos Tunnels. Although Devil's Slide is the name given to a coastal promontory on the San Mateo County coast that is subject to ongoing landslides, the old Route 1 built in 1936 ran just east of the area and was also called Devil's Slide. In 1940, a major landslide destroyed much of the road, and the route was continually subject to major repair and closures due to slides and rock falls. More than half a century later, the road was closed for good when the Tom Lantos Tunnels opened. The tunnels provide a safer route through the dynamic area.

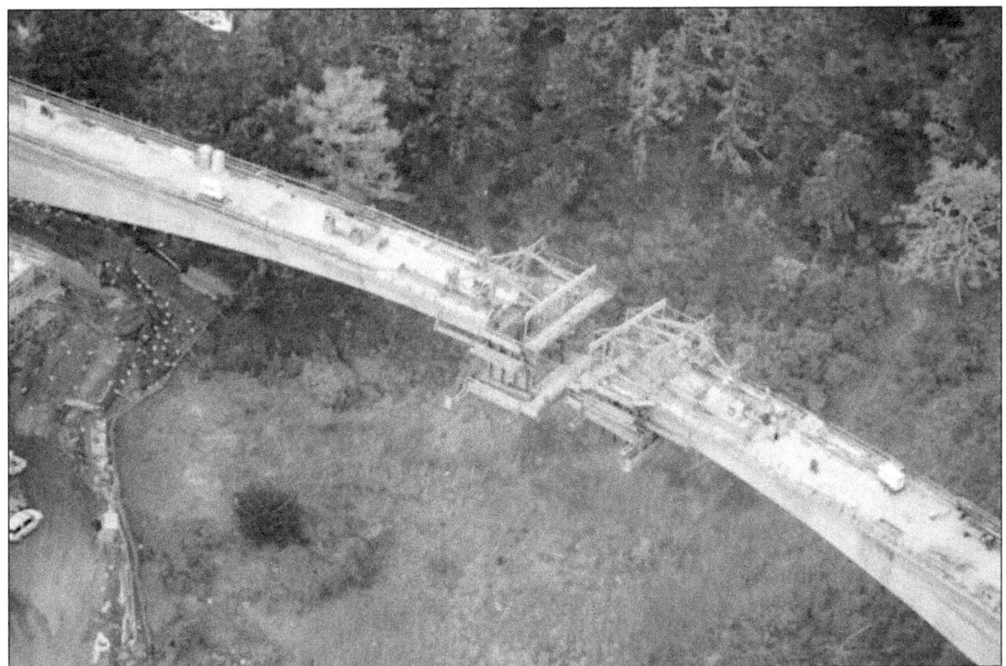

Here is a close-up view of the side-by-side bridges under construction at Devil's Slide. The bridges have a total length of approximately 1,000 feet with two twin piers on each side of the valley and feature main spans of approximately 445 feet.

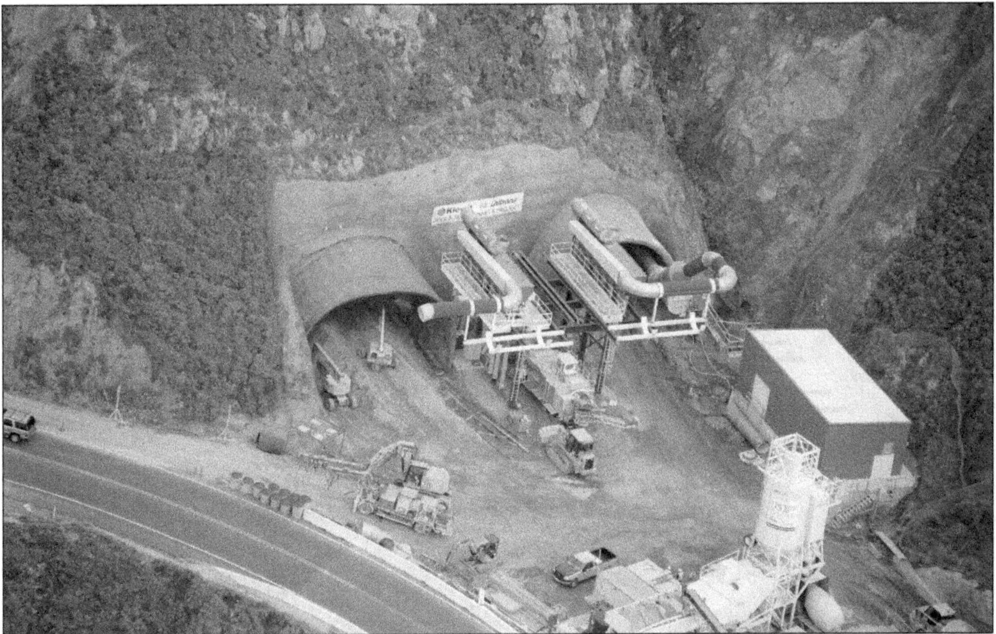

This is an aerial view of the construction of the Tom Lantos Tunnels through San Pedro Mountain as seen in 2009. Ground was broken for the tunnels on May 6, 2005, and boring began on September 17, 2007. The tunnels are each 30 feet wide and 4,200 feet long.

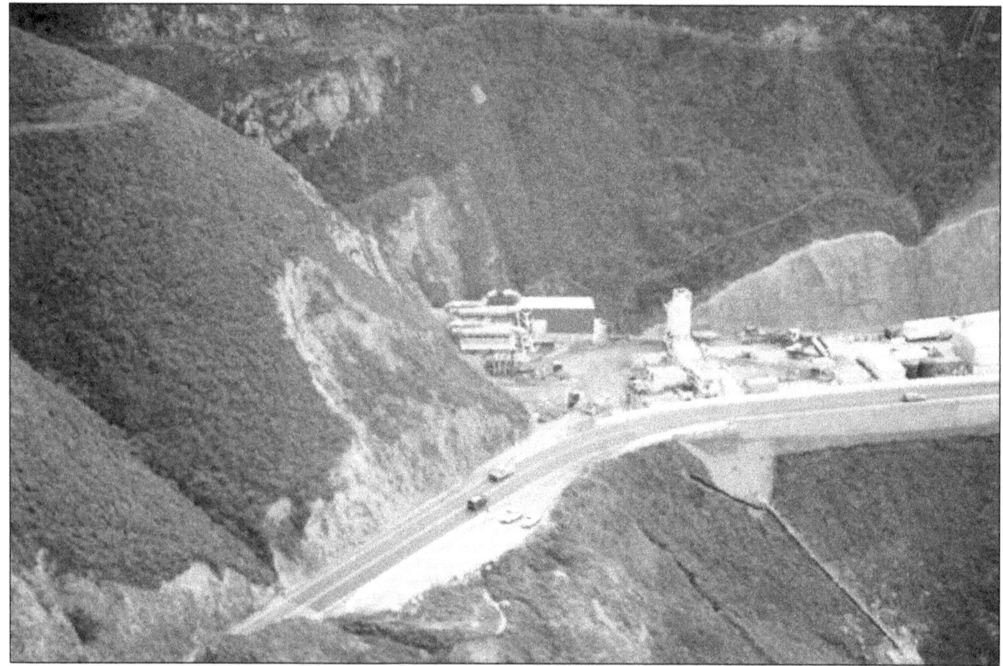

The Tom Lantos Tunnels were determined to be a safe alternative to circumvent the Devil's Slide area, which was often closed from landslides, and extremely dangerous to motorists traveling along that section.

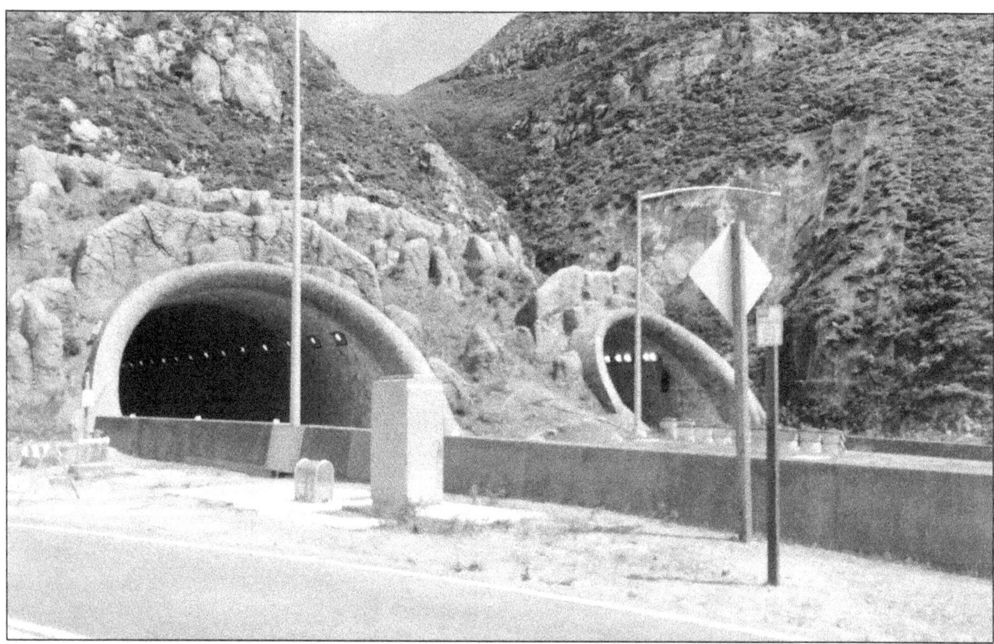

The Tom Lantos Tunnels opened on March 25, 2013. Excavation removed 11.4 million cubic feet of rock. The tunnels run north and south between Pacifica and Montara. They are named after the late congressman Tom Lantos, who played a significant role in acquiring funding for the project.

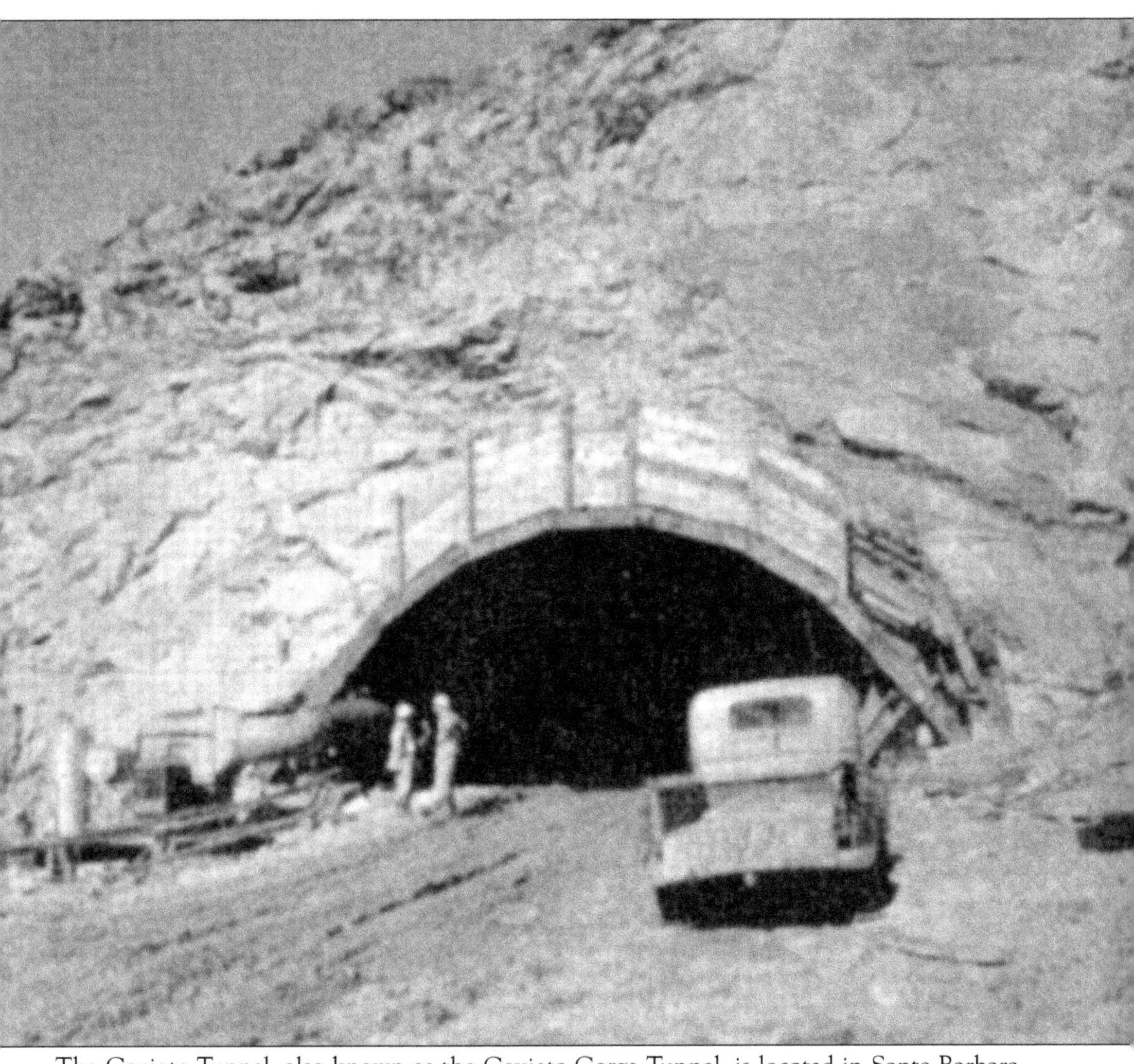

The Gaviota Tunnel, also known as the Gaviota Gorge Tunnel, is located in Santa Barbara County on US Route 101, which runs 21 miles along the Gaviota coast. The northbound tunnel was completed in 1953. Pictured is the exterior of the tunnel during construction in 1950. The tunnel is 420 feet long and 17.5 feet tall. The Gaviota coast is largely undeveloped and is the longest rural coastline remaining in Southern California.

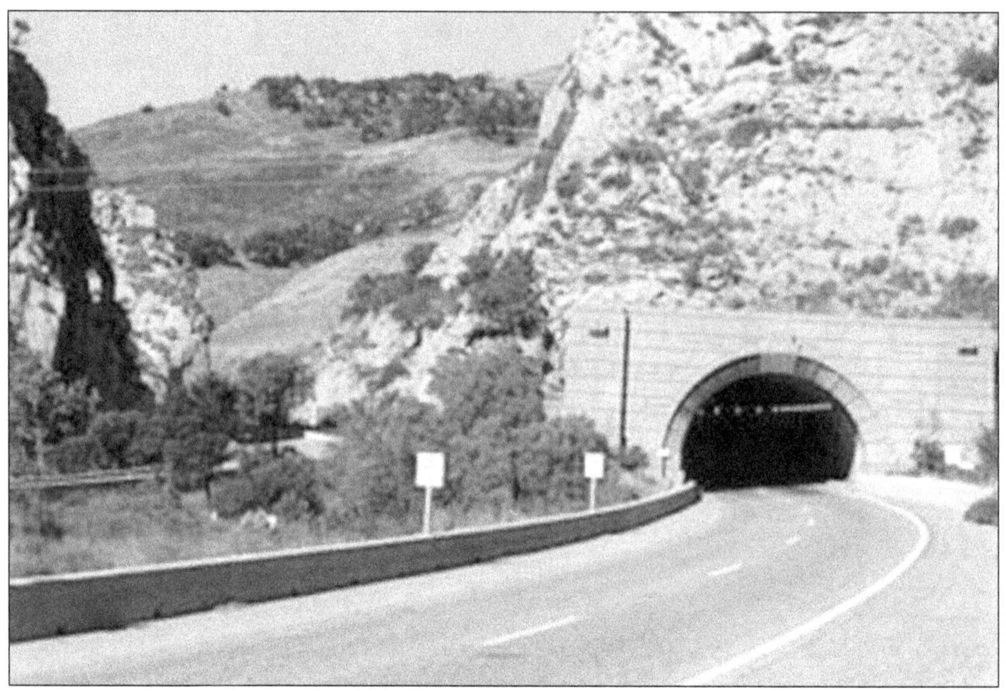

This is the completed Gaviota Tunnel. Prior to the construction of the tunnel, a narrow two-lane north-south road served motorists for decades through the Gaviota Pass. Instead of widening the two-lane road, it was decided that a tunnel would be a better option.

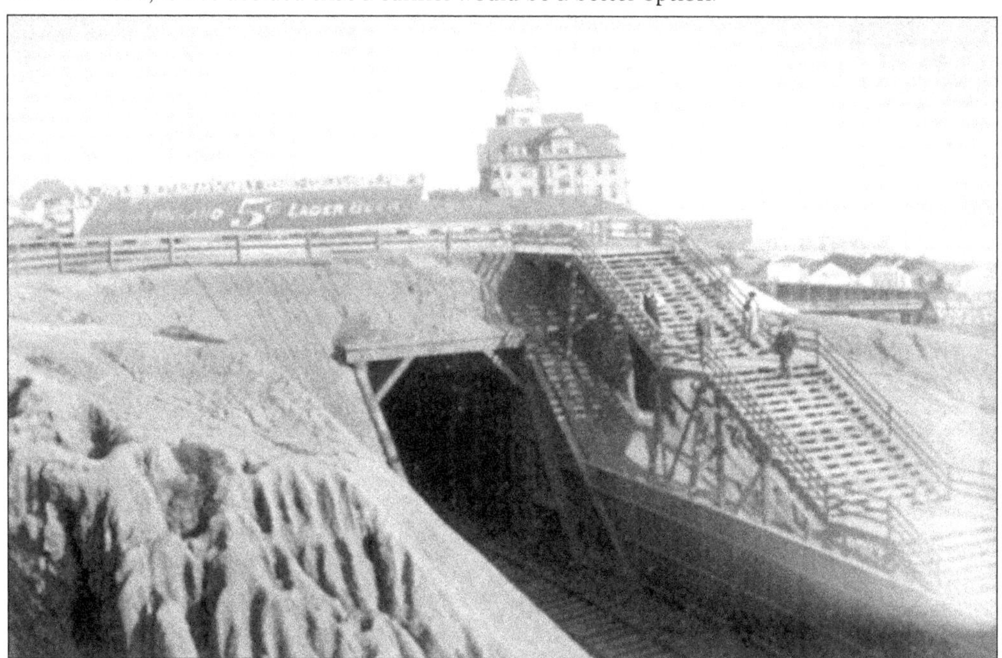

Pictured is construction of a tunnel in Santa Monica in the early 1930s. The tunnel was originally constructed to be a Southern Pacific Railroad tunnel, but it was reconstructed for automobiles and named the Olympic Tunnel. In the background is the Arcadia Hotel. To the right is the famed Santa Monica Pier.

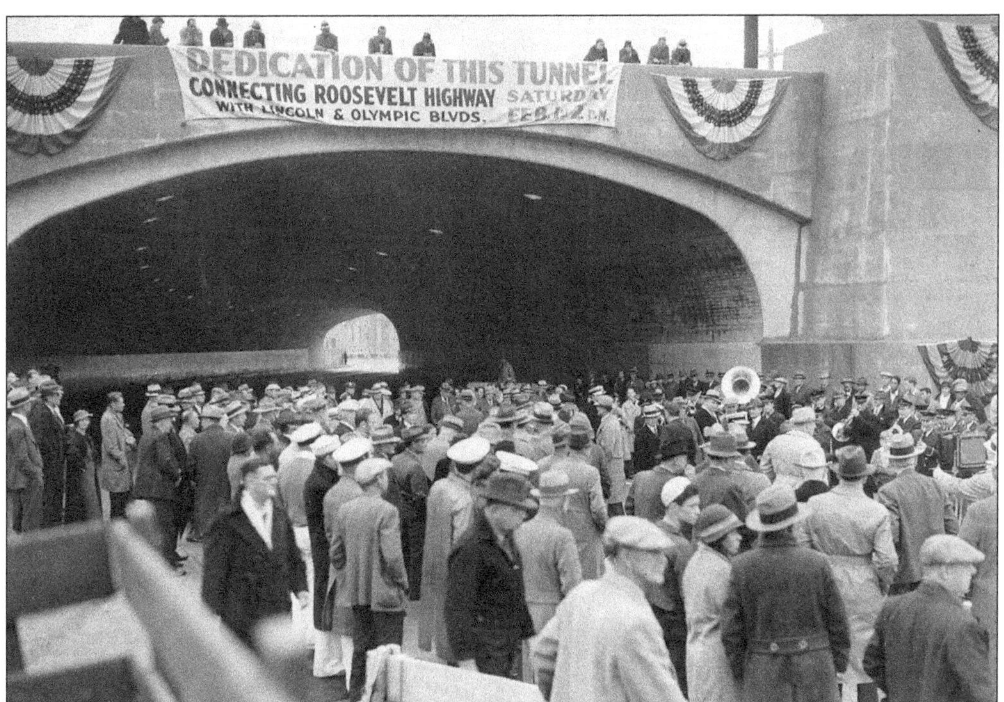

A large crowd gathers in front of and above the tunnel entrance to celebrate the opening of the Olympic Tunnel in Santa Monica. The tunnel links the Roosevelt Highway, later designated Route 1, to Olympic Boulevard and Lincoln Boulevard in the city of Santa Monica. (Bob Jakobsen.)

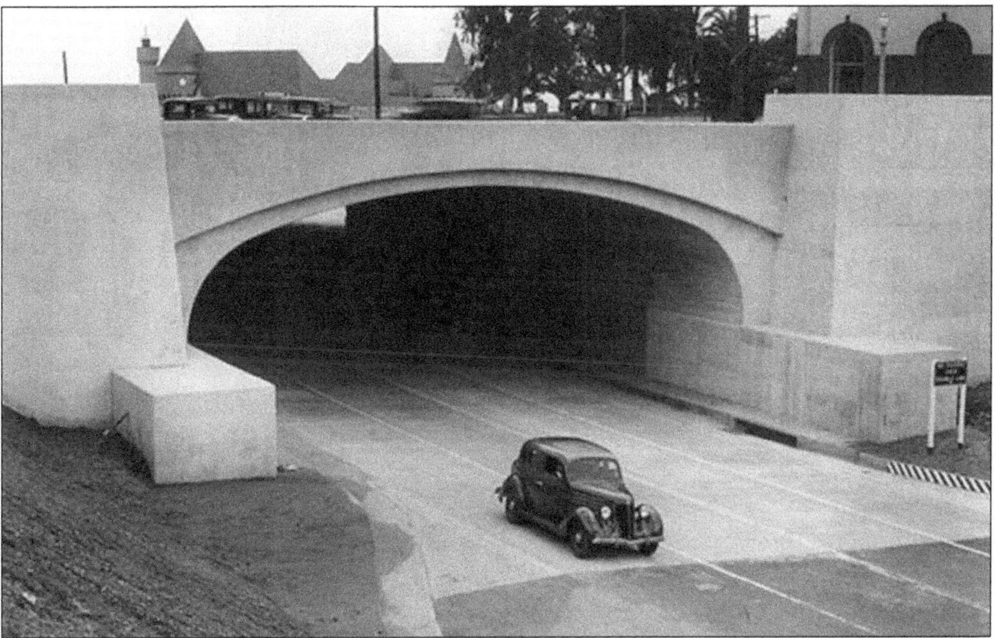

The first automobile drives through the Olympic Tunnel in 1936. The tunnel was renamed in 1969 in honor of Robert E. McClure, editor of the *Santa Monica Outlook* newspaper and advocate for the highway. The length of the tunnel is approximately 400 feet, and it is designated as part of State Route 1. (Bob Jakobsen.)

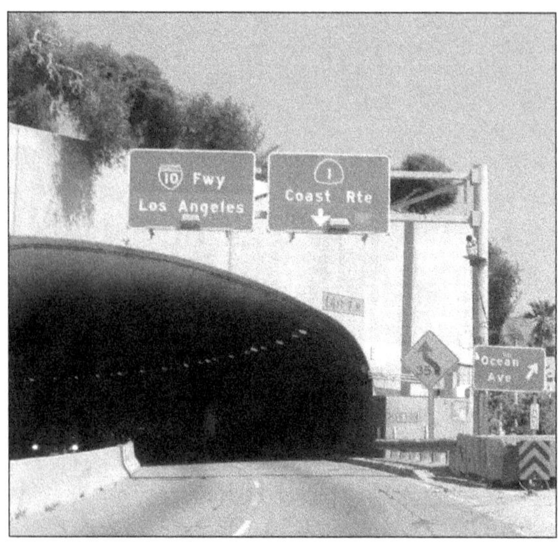

The McClure Tunnel is shown here in 2017. The coast highway from southern Ventura County into Los Angeles County is typically congested for approximately 28 miles, especially during rush hours. Daily traffic crawls past the Santa Monica Pier and through the tunnel. It is estimated that 75,000 vehicles travel the stretch daily.

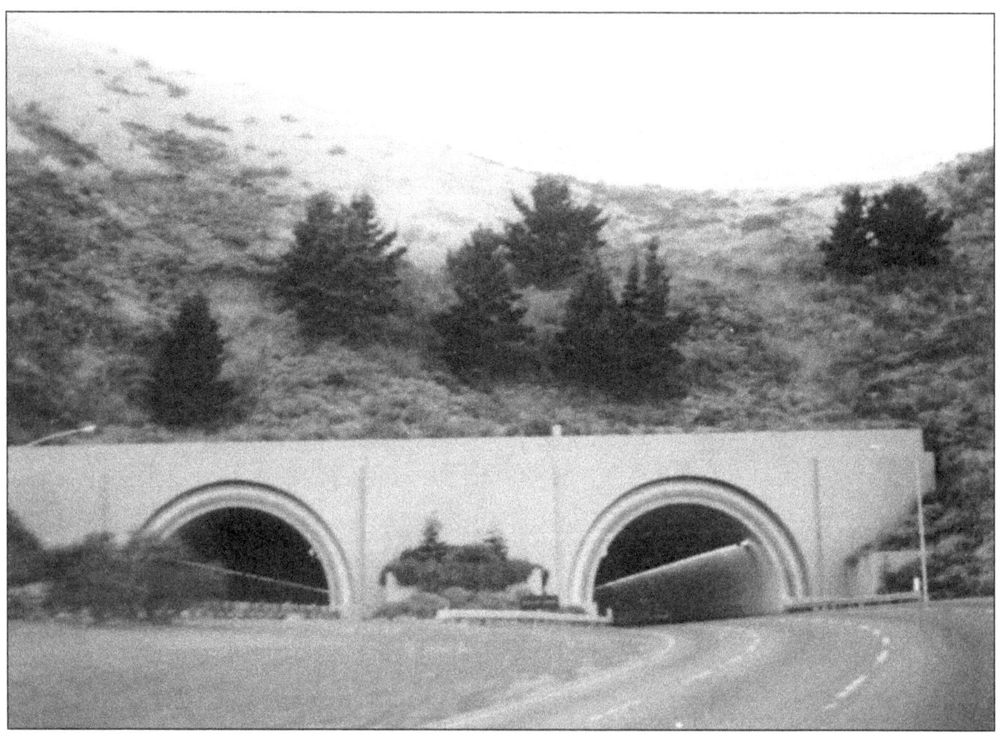

The Waldo Tunnel is pictured in 2000 along Highways 101 and 1 in Sausalito. The first bore of the tunnel was in 1937, and the second bore was in 1954 The tunnel connects the city of San Francisco to Marin County and North Bay. It has been renamed the Robin Williams Tunnel.

Three

CALIFORNIA'S LONGEST NORTH-SOUTH HIGHWAY

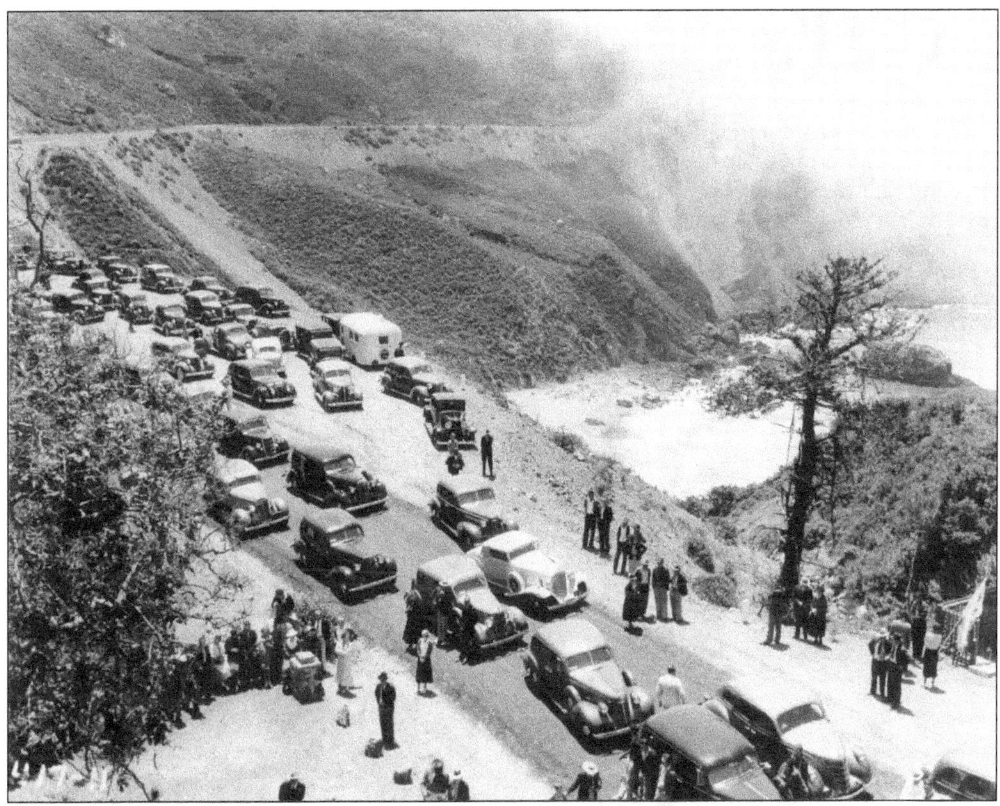

Pictured is a 1937 caravan of over a hundred automobiles that attended the long-awaited celebration of the final section of the coast highway, originally named the Roosevelt Highway, in Big Sur. The opening of the road connected the Big Sur area with the rest of California. Motorists were now able to travel the entire California coast. The highway was designated California's first State Scenic Highway and has since been protected by the Highway Beautification Act of 1965.

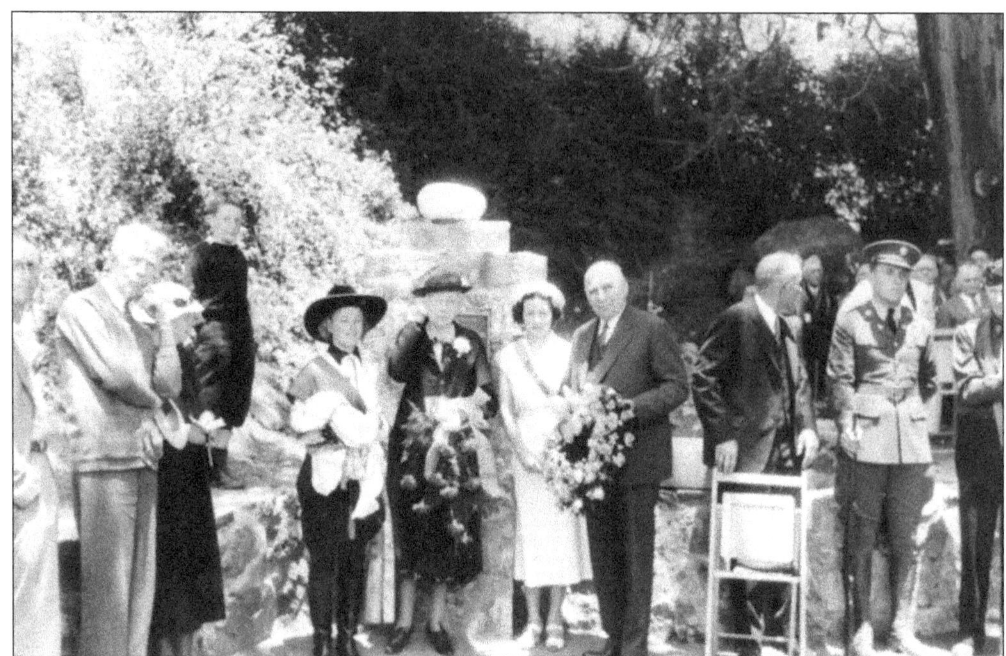

Holding a bouquet wreath is California governor Frank F. Merriam at the 1937 dedication of the Elmer Rigdon Memorial Drinking Fountain, located four miles north of Lucia. State senator Rigdon of Cambria was instrumental in acquiring state appropriation for the Carmel to San Simeon Highway. Below, Governor Merriam is at the opening of the Carmel to San Simeon Highway ribbon-cutting ceremony in 1937. Hundreds of people attended the event, including Dr. John L.D. Roberts from Monterey, who first conceived of a coast road while traveling along the old Big Sur wagon road. Standing at left is Joyce Matheson, Miss Cambria Pines of 1937. (Both,)

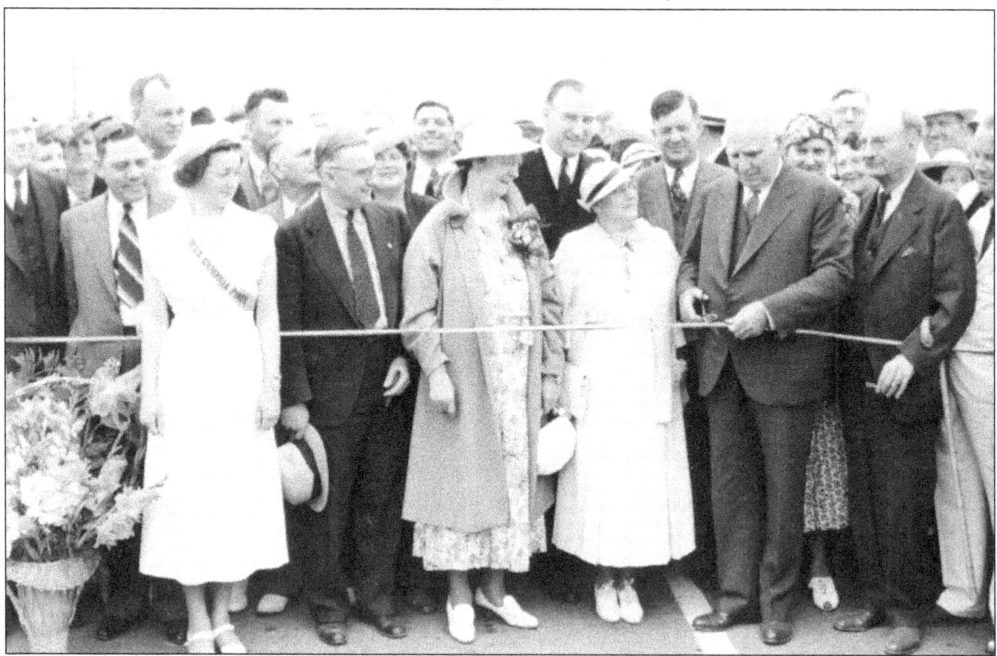

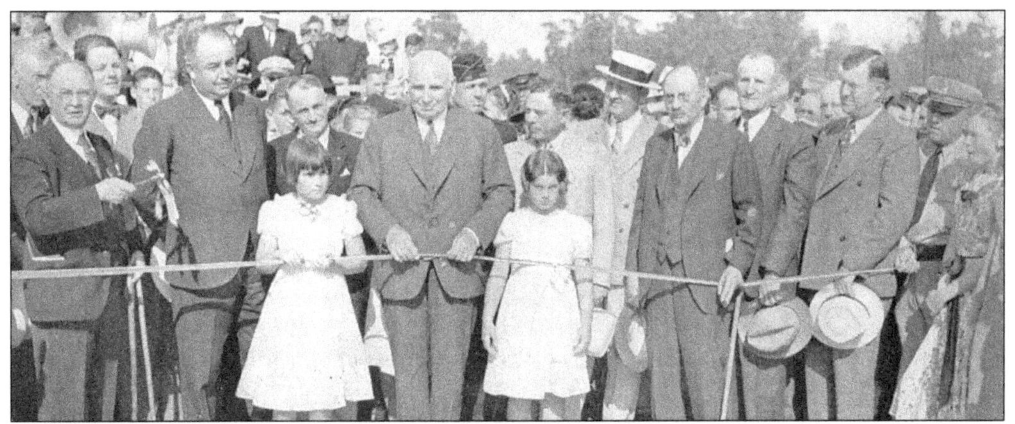

Pictured above is the 1937 ribbon-cutting ceremony of the Roosevelt Highway between Santa Monica and Seal Beach in Southern California. From left to right are councilman Franklin Buyer, chairman of the day; E.J. Amar, president of Los Angeles Harbor Commissioners; state highway commissioner W.T. Hart; California governor Frank F. Merriam; Harry Hopkins, state highway commission chairman; supervisor Leland Ford, Los Angeles; highway commissioner Phil A. Stanton; district highway engineer S.V. Cortelyou; and assistant director of public works Justice Craemer. Governor Merriam cuts the ribbon, opening the new road to traffic. Below is another photograph of the dedication ceremony showing a portion of the crowd that attended the celebration. (Below, Los Angeles County Metropolitan Transportation Authority.)

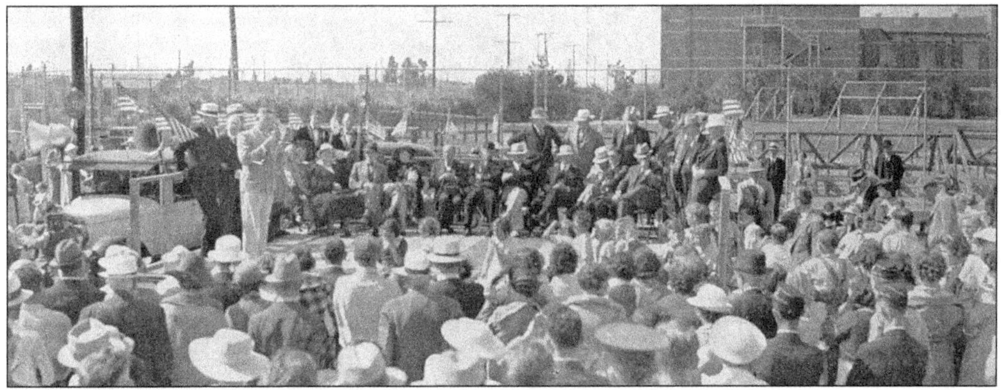

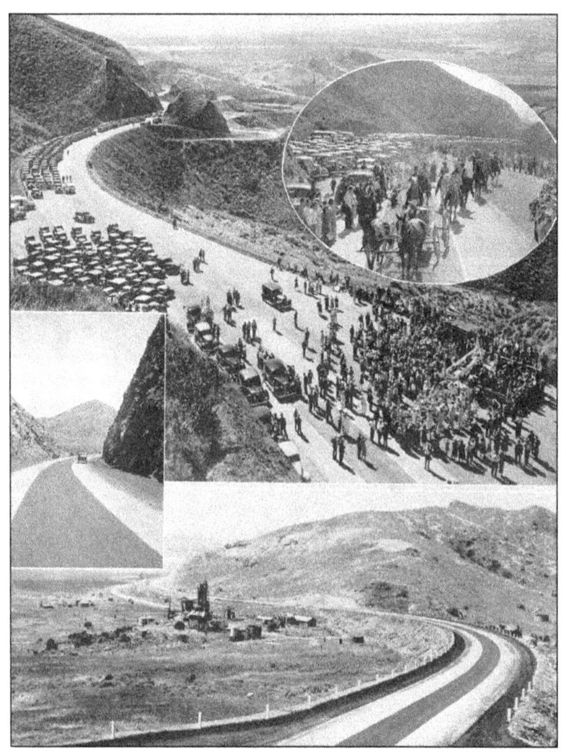

The 1937 collage at left by California Highways and Public Works shows the dedication of the Conejo Grade highway in Ventura County. A small parade was held for the event. A portion of the transportation pageant is shown, with riders on a buggy and several people following on horseback. A large crowd celebrated the new alignment of the grade. At left center is the old Conejo Grade road, and at bottom is a view of the new highway looking north toward Camarillo. Below, hundreds of automobiles gather at Cambria Pines Lodge to caravan to San Simeon for the dedication of the completed San Simeon to Carmel highway on June 27, 1937. Gov. Frank Merriam led the caravan in the first car. A boulder was pushed off the road, symbolically opening the highway to traffic. (Left, Los Angeles County Metropolitan Transportation Authority; below, copyright 2007 California Department of Transportation. All rights reserved.)

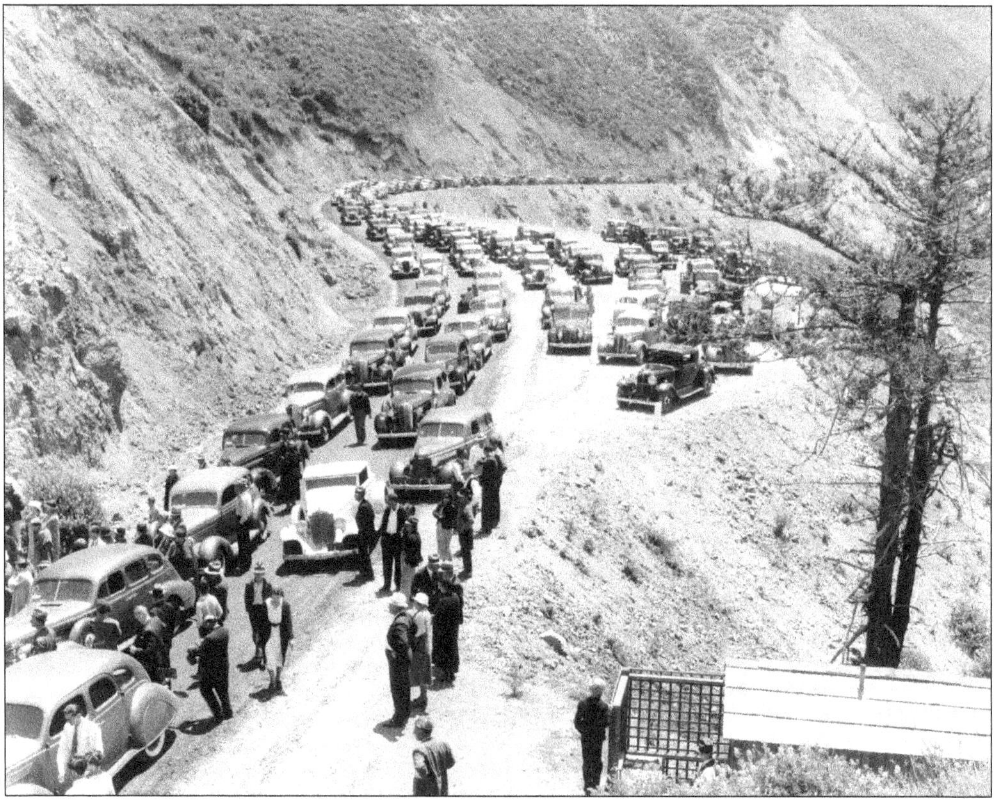

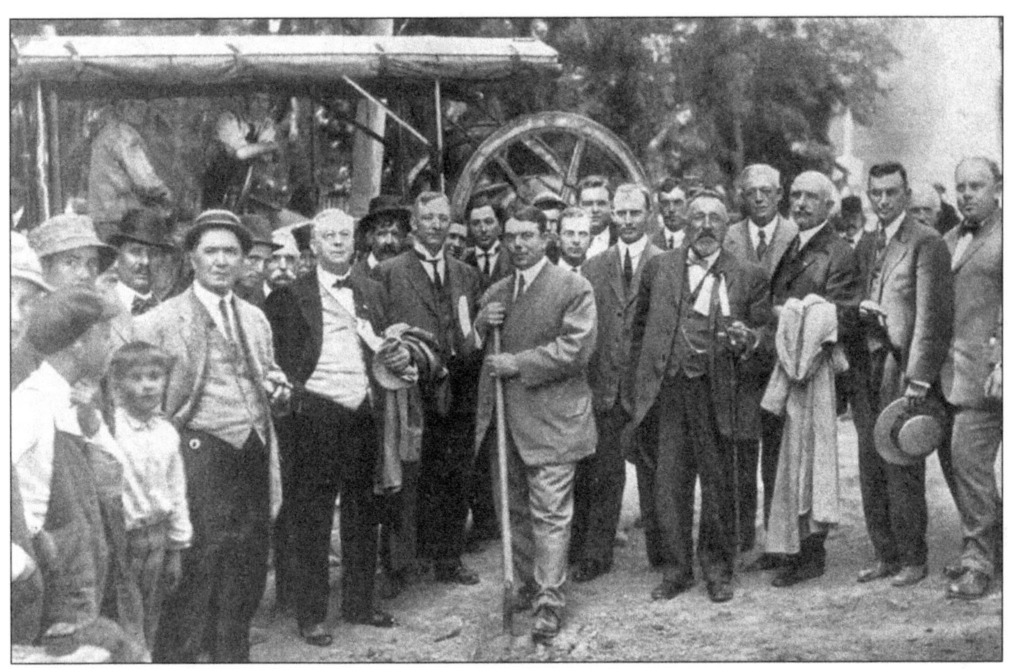

In 1912, Burton A. Towne, chairman of the Highway Commission, turned the first shovelful of earth on the coast highway in San Mateo. The first state highway construction began at this location. Contract No. 1 was the first contract awarded by the California State Highway Commission. Pictured above is the first section of highway under construction in San Mateo County, which was located between San Francisco and Burlingame. This photograph was featured in the book *California Highway* published in 1920. The section of highway is pictured below after construction in the 1920s. (Below, Los Angeles County Metropolitan Transportation Authority.)

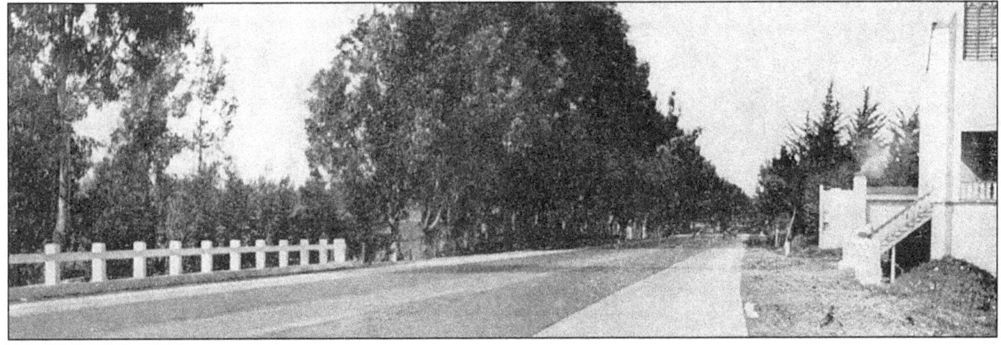

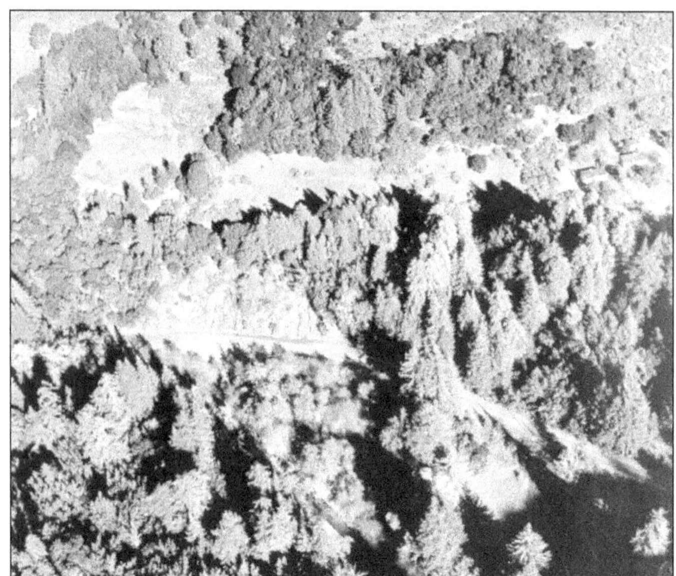

The topography of the Big Sur wilderness area is characterized by steep hillsides and sharp-crested ridges. Dense chaparral blanket the hills along with coast live oak, valley oak, and several varieties of pine. Highway 1 wends for miles through the scenic green landscape. (Copyright California Department of Transportation. All rights reserved.)

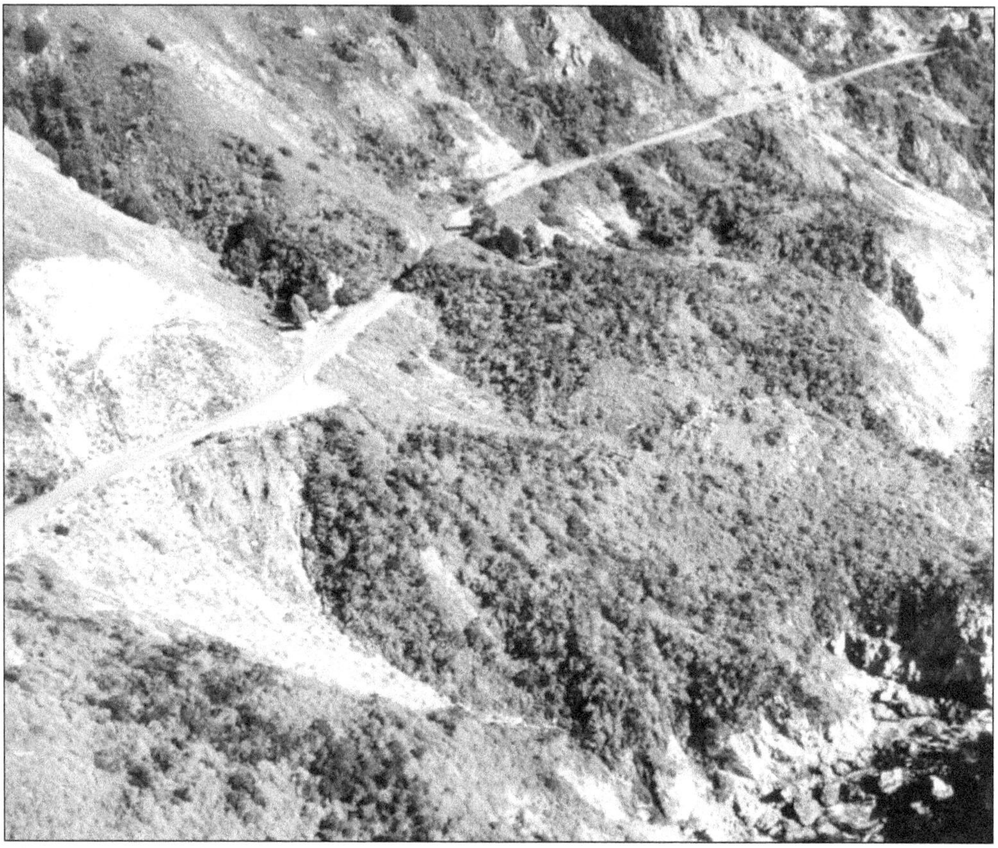

This is a 1963 aerial view of the highway emerging from the dense redwood forest. The coast road to Cambria from this point is called "the hundred-mile drive." It is one of the most scenic sections of the highway, with the road perched between steep slopes and near-vertical cliffs. (Copyright California Department of Transportation. All rights reserved.)

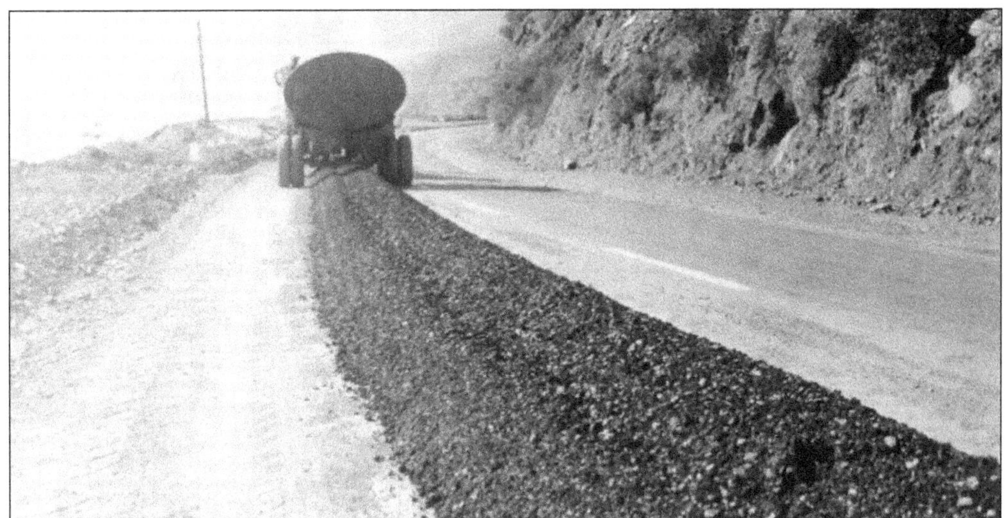

A freshly graveled road along the coast in Monterey is seen in 1932. Preservation of the coast road requires continued maintenance to lessen the effects of natural erosion and harsh coastal elements. (Copyright California Department of Transportation. All rights reserved.)

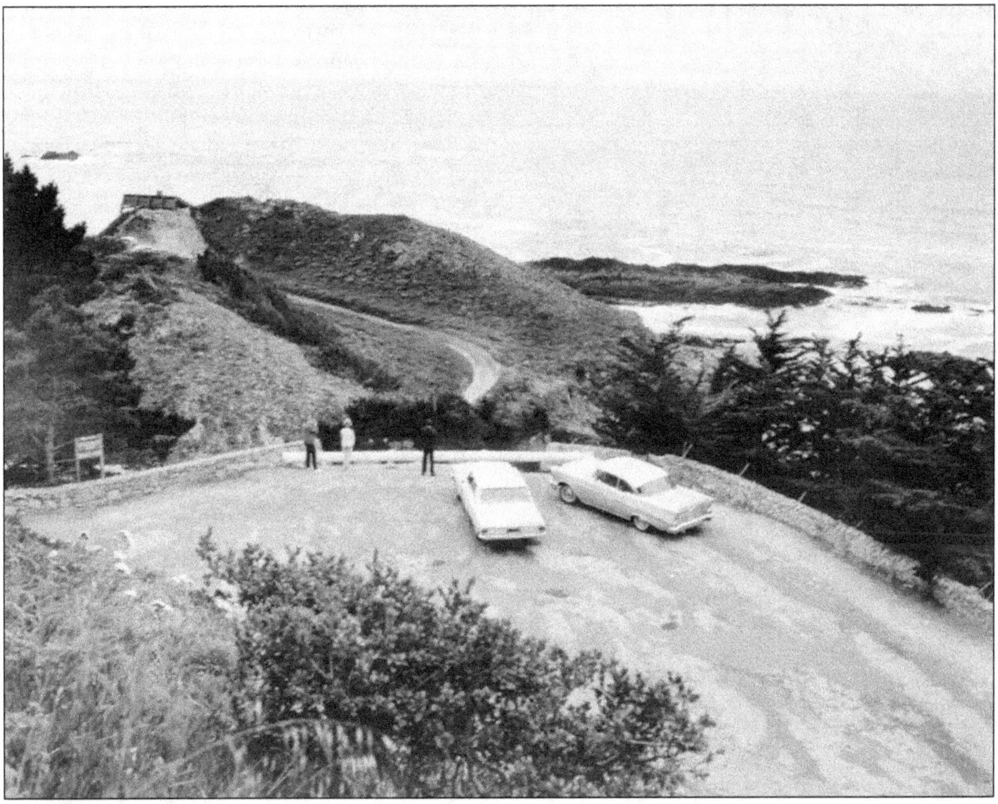

It has been said that the coast road was built to see the magnificent Big Sur Coast. This roadside scenic view along the Monterey coast is pictured in 1965. The panoramic, unobstructed views from designated view points are the safe way to take in the splendor. (Copyright California Department of Transportation. All rights reserved.)

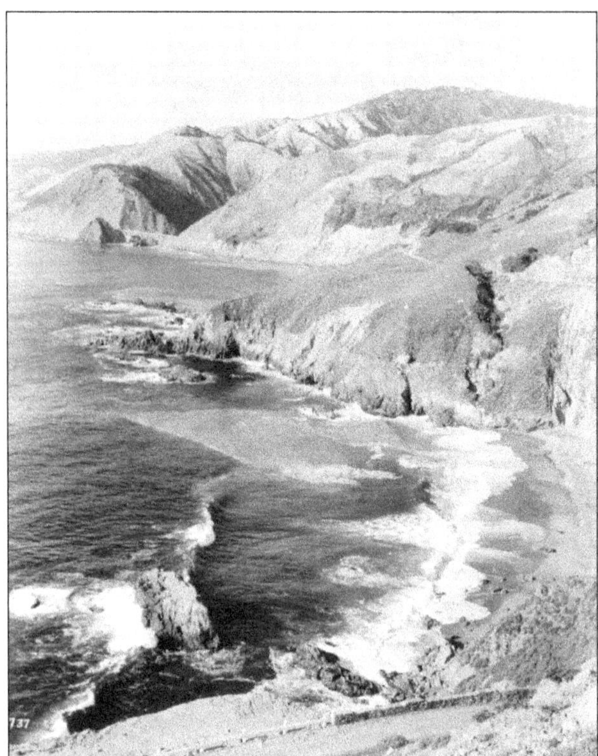

The Big Sur Coast is seen here in 1940. The coast is largely comprised of granite, marble, and sandstone. Its beauty and splendor can be seen and appreciated from the highway, especially at McWay Creek, where its waterfalls pour into the ocean at Julia Pfeiffer Burns State Park. (Copyright California Department of Transportation. All rights reserved.)

A two-lane Highway 1 runs along the edge of steep cliffs on the Big Sur Coast. Road cuts can increase erosion and weaken hillsides. Rain, storms, and earthquakes can cause rock and earth to fall on the highway at any time of the day or night. (Copyright 1963 California Department of Transportation. All rights reserved.)

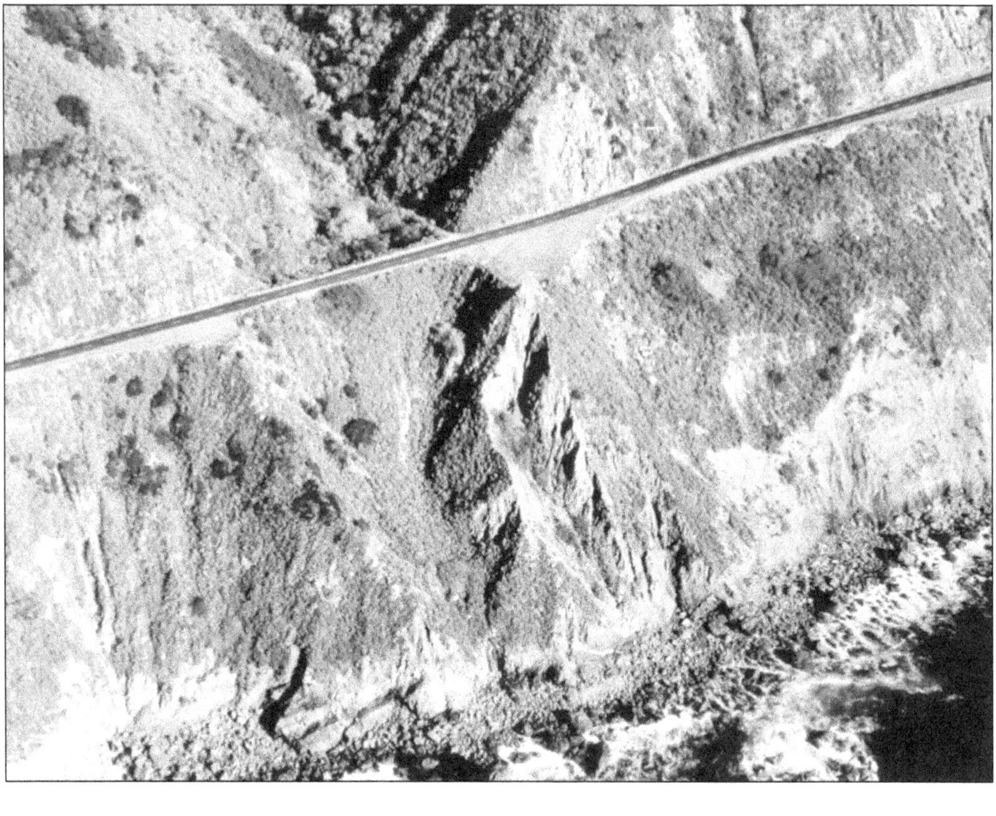

This is an aerial view of Highway 1 following along the steep cliffs of the Monterey coast. The dramatic and unobstructed panoramic views of the jagged coastline are breathtaking. This section requires regular maintenance due to erosion, strong storms, wind, and heavy rain. (Copyright California Department of Transportation. All rights reserved.)

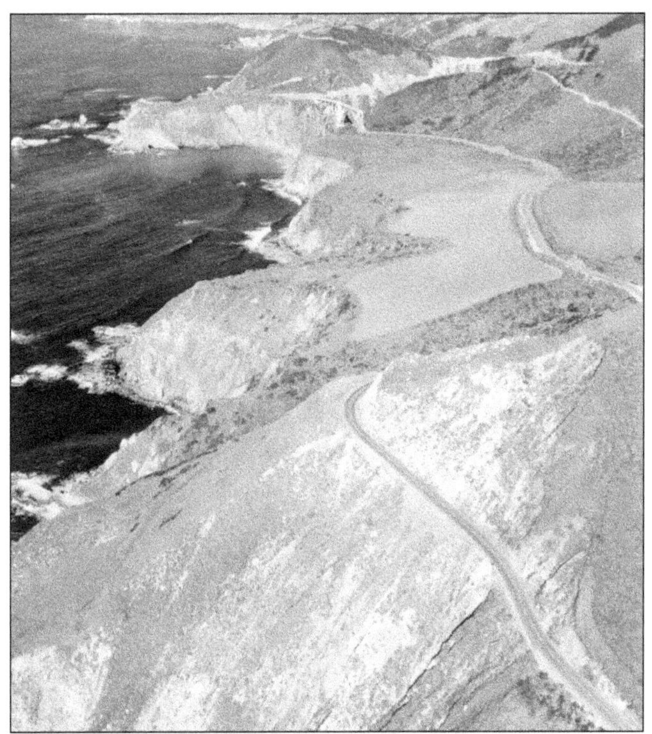

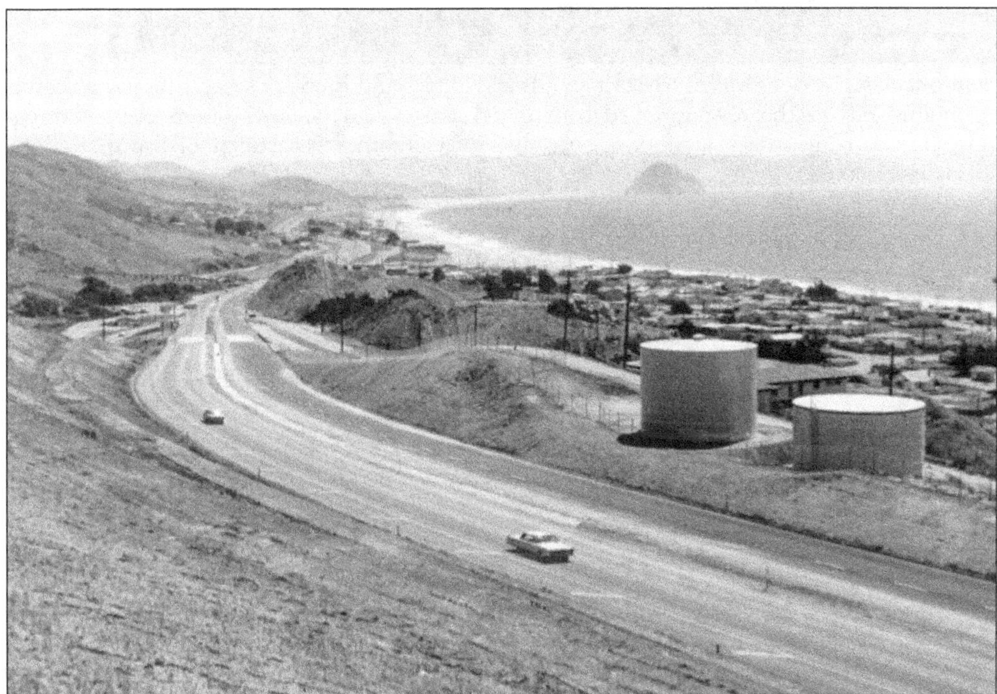

Highway 1 along the San Luis Obispo coast enters Morro Bay on the Cayucos Bypass. Morro Rock, visible in the distance, is an offshore stack connected to the mainland by a tombolo, a narrow piece of land that joins an island to the mainland. (Copyright 1966 California Department of Transportation. All rights reserved.)

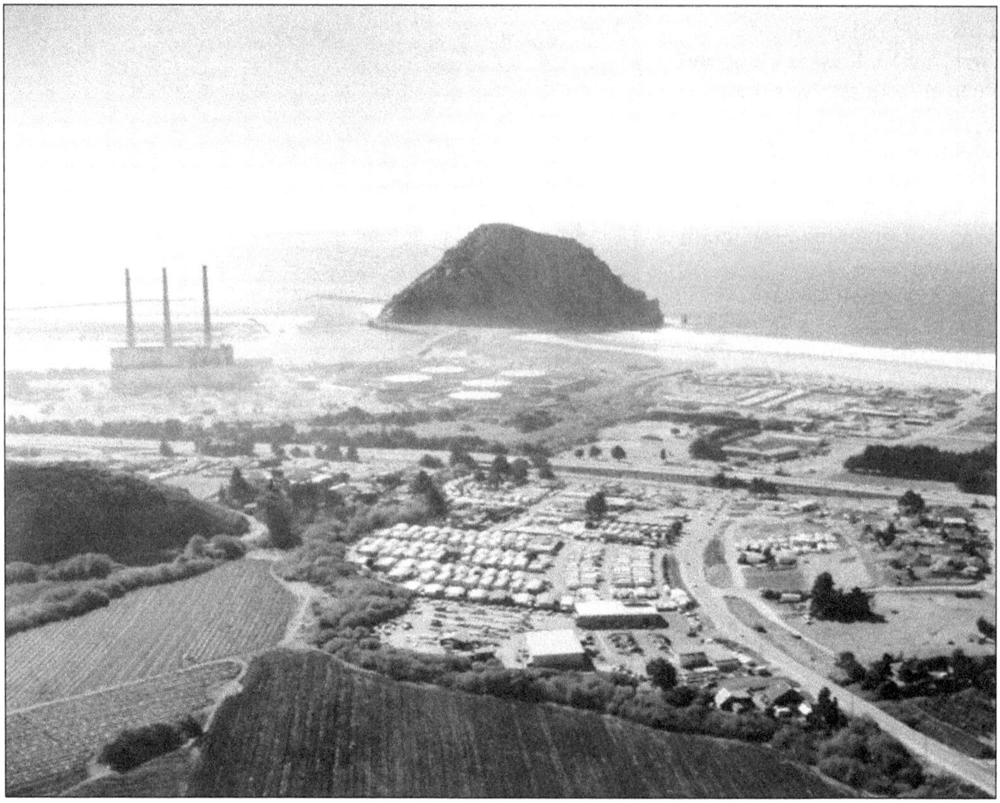

Here is a closer view of Morro Rock. It stands at 576 feet and is one of several similar volcanic mountains and hills that are positioned in an inland chain called the Seven Sisters, or the Morros, in western San Luis Obispo County.

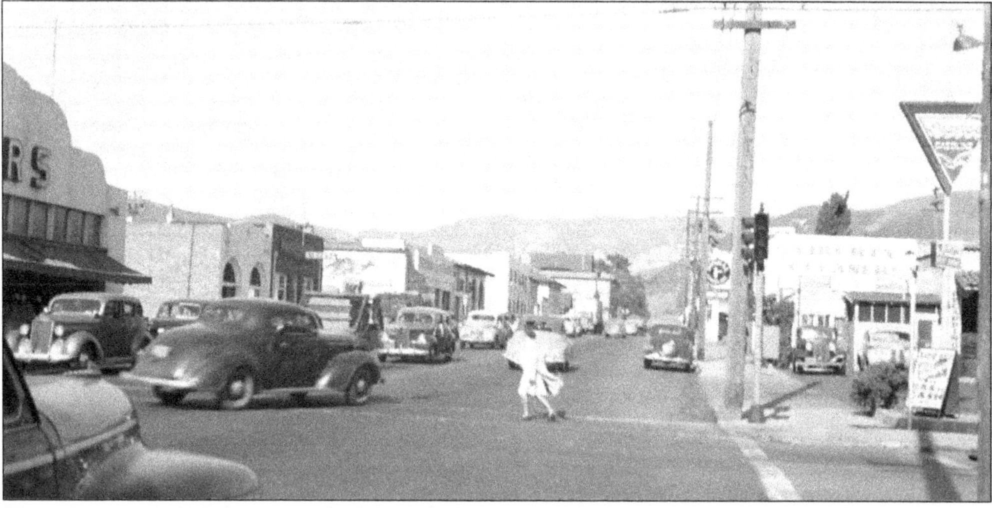

This 1947 view of Highway 101 looks southeast toward the Broad and Marsh Streets intersection in San Luis Obispo. It appears that the pedestrian is either crossing against a red light or the car in the intersection is turning on a red light.

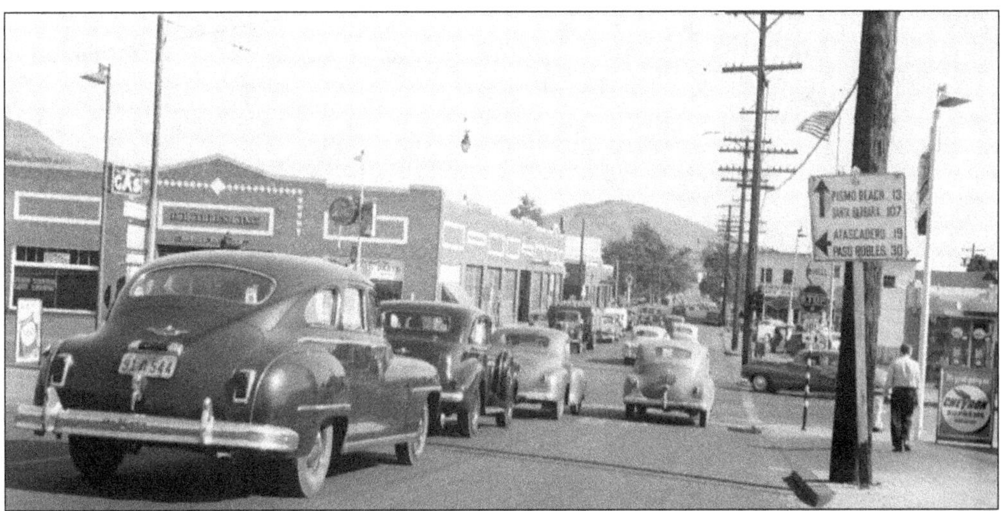

This photograph looks southeast toward the Monterey and Santa Rosa Streets intersection in San Luis Obispo, which is located on US Route 101. Framed by Santa Lucia Mountain east of the city, it is a seismically active area near several faults, including the infamous San Andreas Fault.

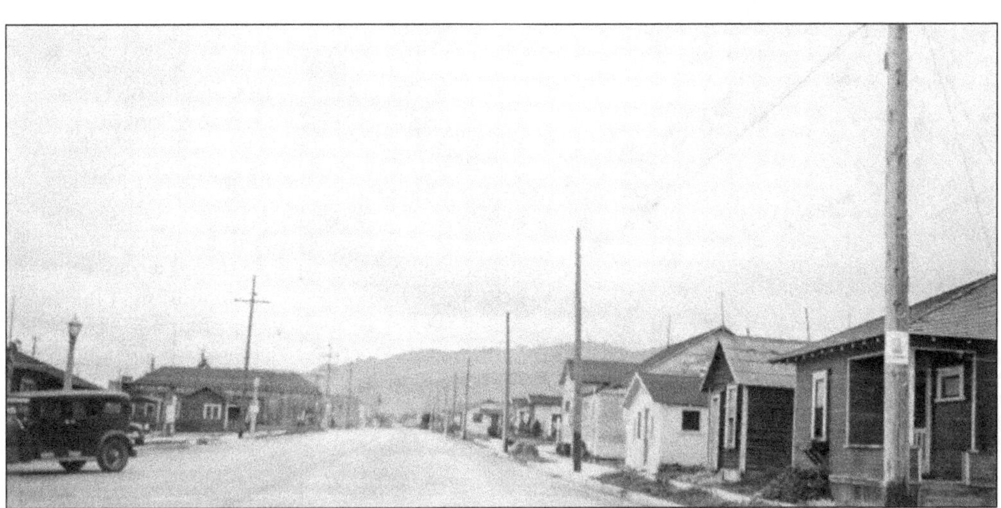
Pictured is North Pismo Beach in 1934. It is one of the Five Cities that include the coastal communities of Grover Beach, Arroyo Grande, Oceano, and Shell Beach. Pismo Creek runs into the Pacific Ocean at Pismo Beach. In 1969, a 100-year storm hit the central coast, flooding the Pismo Beach community.

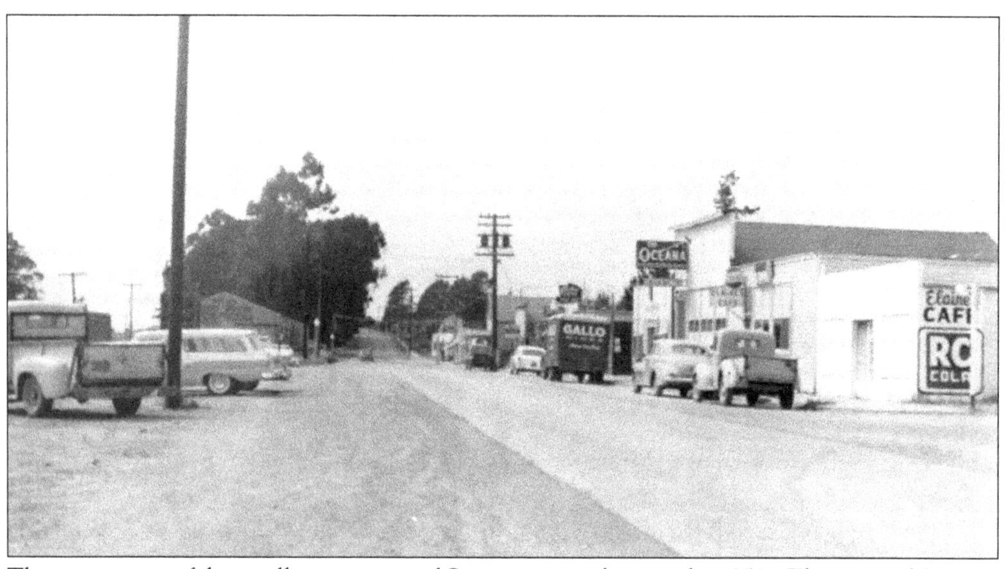

The main street of the small community of Oceano is seen here in the 1950s. The town of Oceano comprises a total area of 1.5 square miles. The town has been plagued by flooding over the years. The 100-year storm in 1969 washed out many coast town roads.

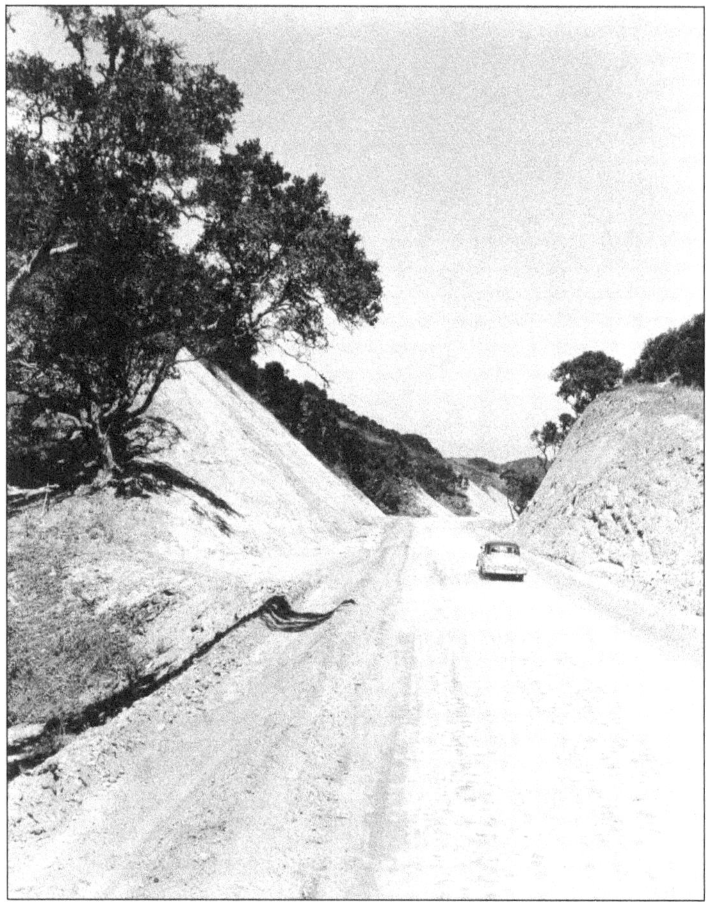

The coast highway through Monterey is seen here looking north. The road was under maintenance by the application of a chip and seal process. Pavement is sprayed with a layer of AC emulsion, a mixture of asphalt and positively charged emulsions, and coated with gravel. The emulsified asphalt binds the gravel to the existing pavement, and vehicle traffic compacts the surface.

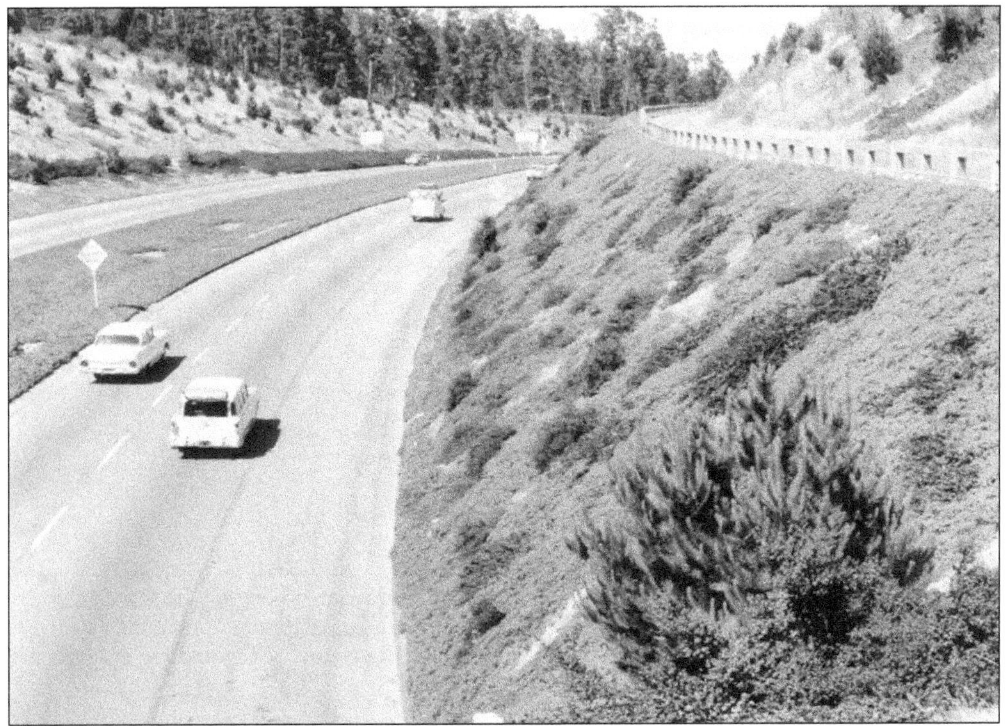

Pictured are northbound and southbound lanes through Monterey. Medians and barriers have changed over the years, incorporating safety and aesthetics into highway design. Monterey Bay is to the north. The southbound road, known as the 17-Mile Drive, leads to the community of Pebble Beach. (Copyright 1947 California Department of Transportation. All rights reserved.)

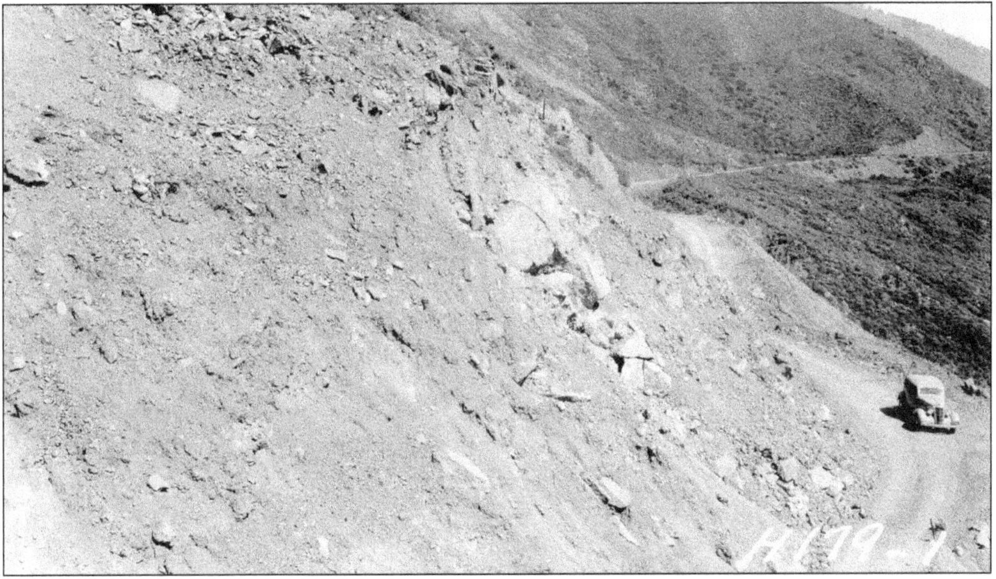

Unstable sloped mountain hillsides are common along the Northern California coast. Falling rock and debris is a recurring maintenance problem for the highway today, even more so in the early years before protective netting and barrier walls were available to hold back unstable hillsides. (Copyright 1937 California Department of Transportation. All rights reserved.)

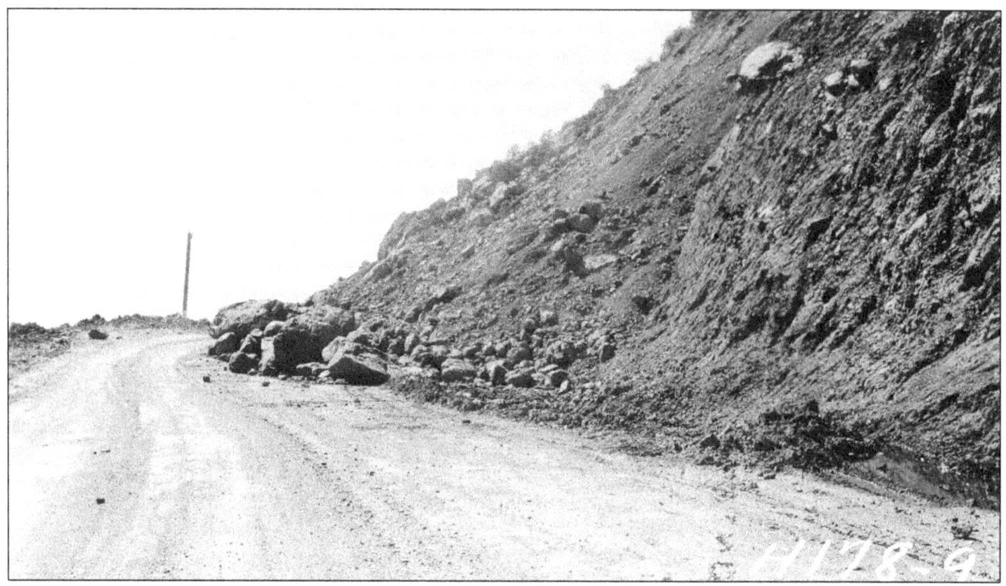

A landslide along the Northern California coast is seen here. Large boulders have rolled down an unstable hillside onto the highway, obstructing the road. Landslides are inevitable along the coast due to its dynamic environment, powerful weather events, and early roadway excavations that inadvertently destabilized hillsides.

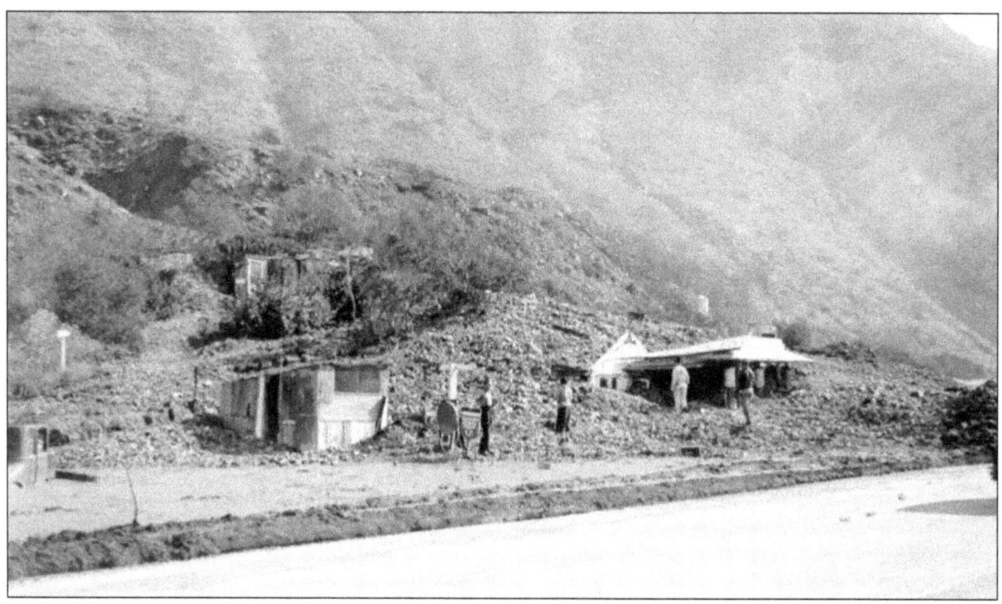

Pictured is a 1933 landslide in Point Mugu. People stand in the rubble of fallen rock that destroyed the Point Mugu service station. The service station was located below the mountain, where natural erosion and the effects of excavation in the building of the highway created an unstable situation.

Erosion along the Ventura coast highway is pictured in 1940. Beach sand protects the coast from erosion by dissipating wave energy and reducing wave damage to cliffs and roads. Inland development and increasing urbanization along the coast interrupt the needed supply of sand transported from upstream mountains to the beaches and the ocean. (Copyright 1940 California Department of Transportation. All rights reserved.)

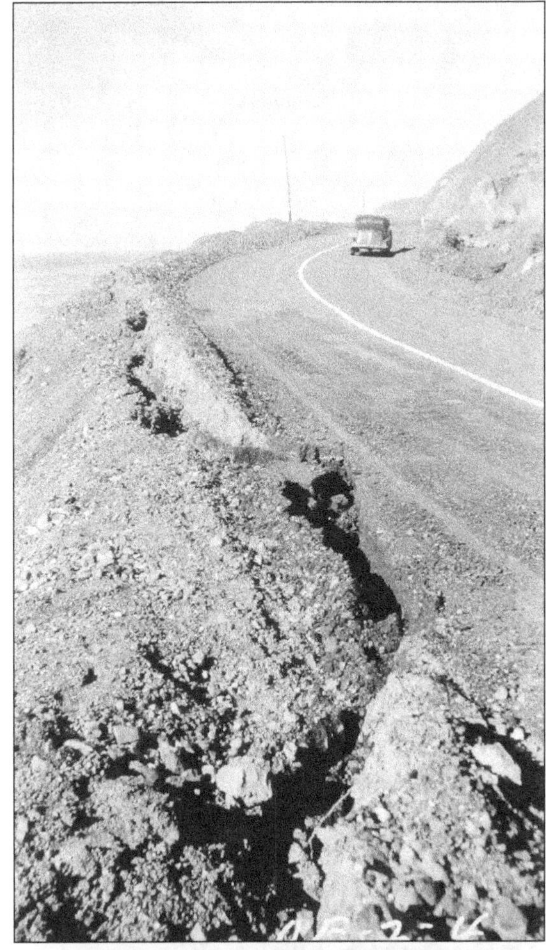

A rural dirt road in Lafler Canyon is seen in 1934 after a flood washed out a large section of the road.

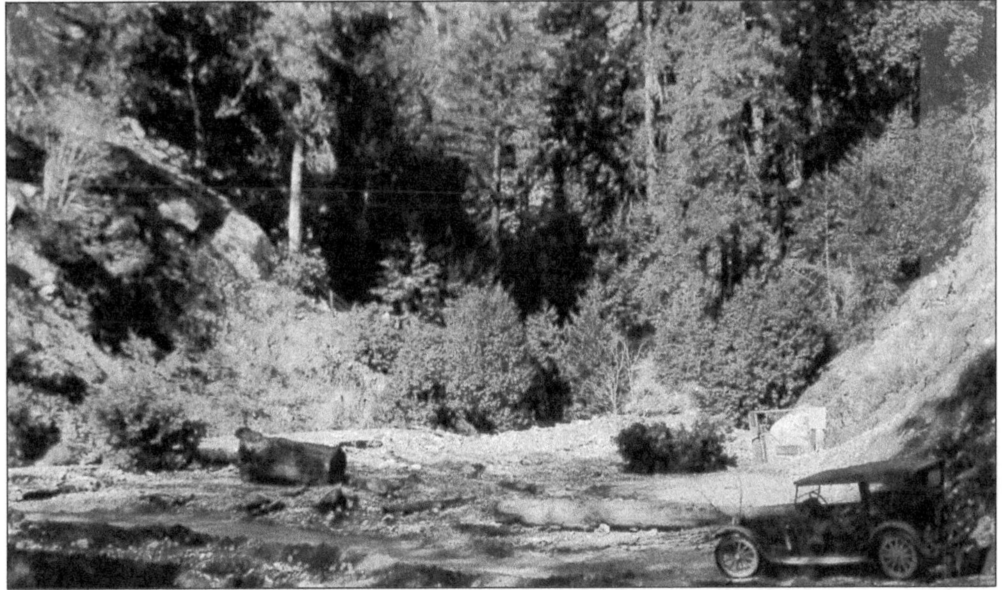

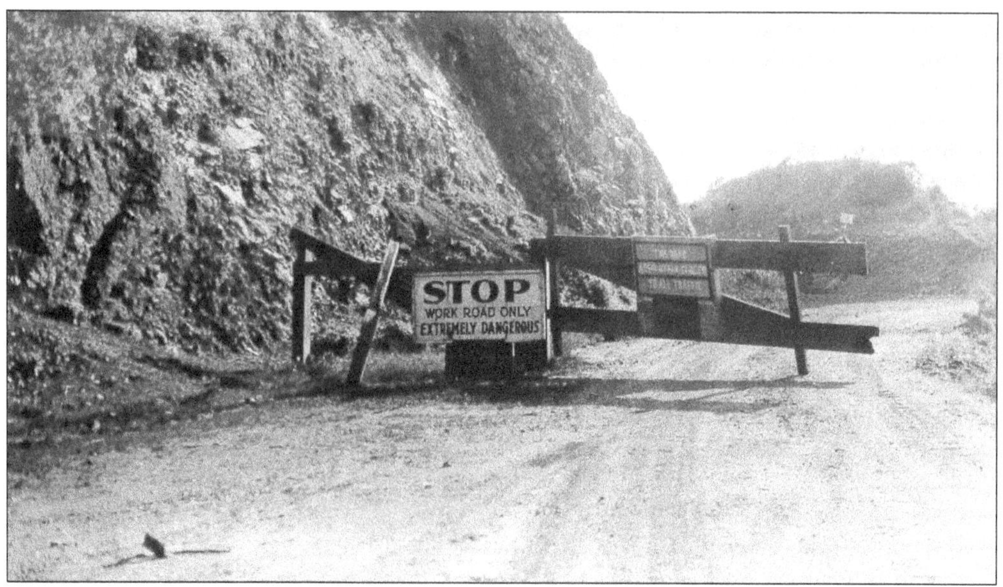

Road closures along California's coast are a frequent sight, particularly after a storm or heavy rain. Most closures occur during the winter months due to harsh weather and dangerous driving conditions along unstable hillsides, yet the coast road is open year-round.

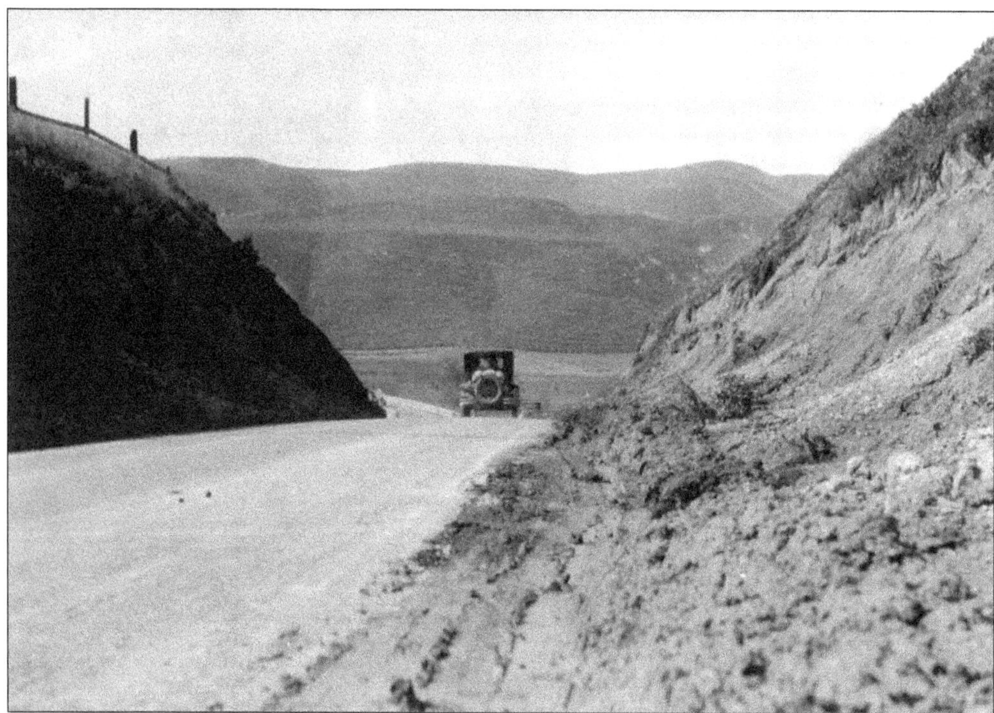

US Highway 101 through Refugio Beach in Santa Barbara County is seen here in 1927. Refugio Beach is framed by the picturesque landscape of the Santa Ynez Mountains. The highway curves along the coast at sea level. At times, flooding has severely damaged or washed out the road.

Highway workers in Santa Barbara inspect fissures and cracks along the highway shoulder from the 1925 Santa Barbara earthquake. A distinct drop between the road surface and the shoulder compromised the integrity of the road.

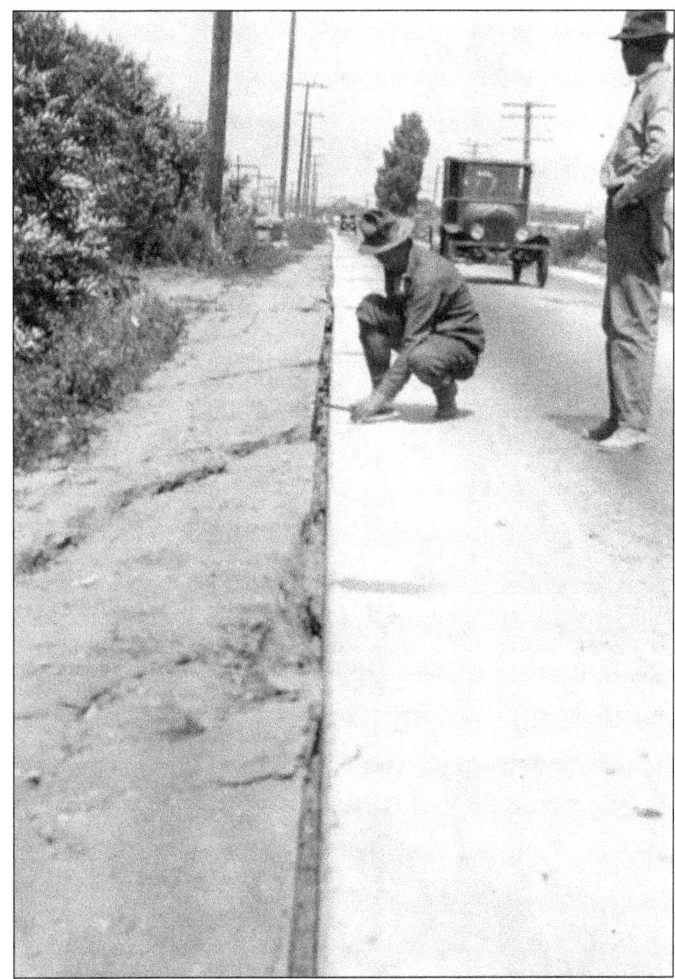

Rocks broke loose from a hillside and spilled onto the highway in Santa Barbara during the 1925 earthquake. California has a long history of earthquakes, including the one with the highest magnitude, 7.8, which hit the San Francisco Bay area in 1906.

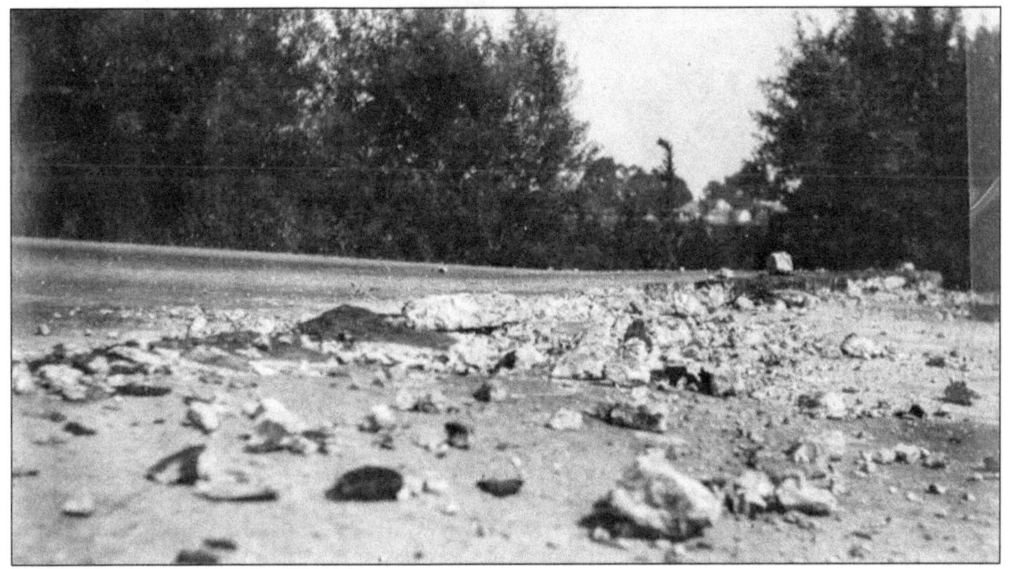

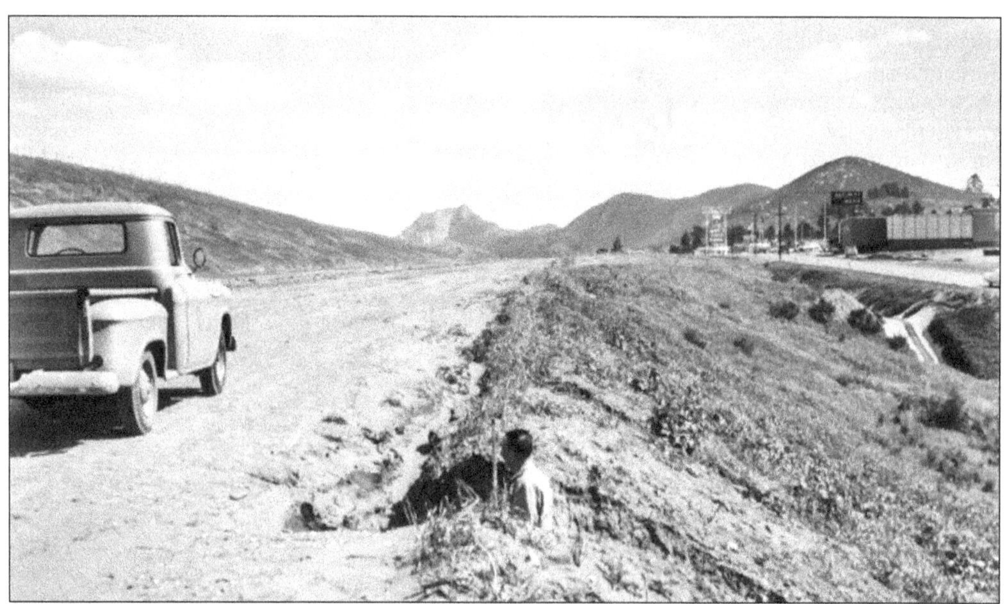

A highway worker in San Luis Obispo County performs storm drainage repair in 1962. Water damage on roads is just as serious and damaging as landslides. It not only threatens adjacent property, but can also weaken the base and subgrade from saturation and ponding for long periods of time.

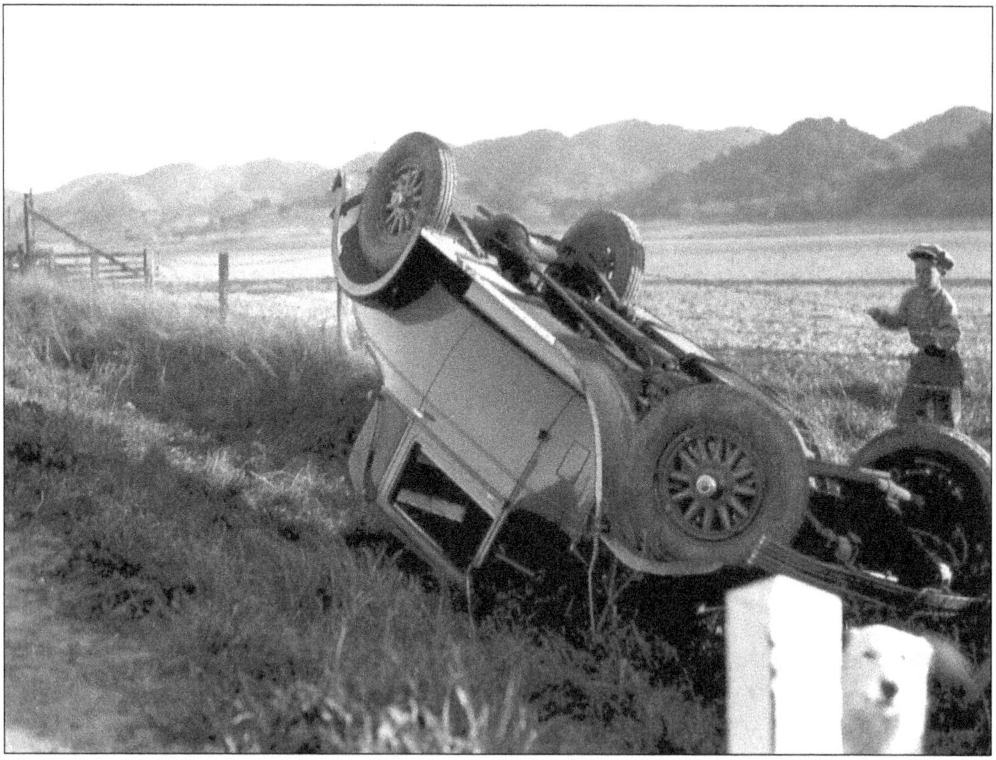

Narrow roads, sharp turns, poorly painted lines, shoulder drop-offs of more than two inches, wheel ruts, potholes, and speed, among other factors, can all cause accidents. This 1928 photograph shows a Chrysler that ran off the road and overturned in a Santa Barbara ditch.

Four

A Nostalgic Road Trip along California's Pacific Coast Highway

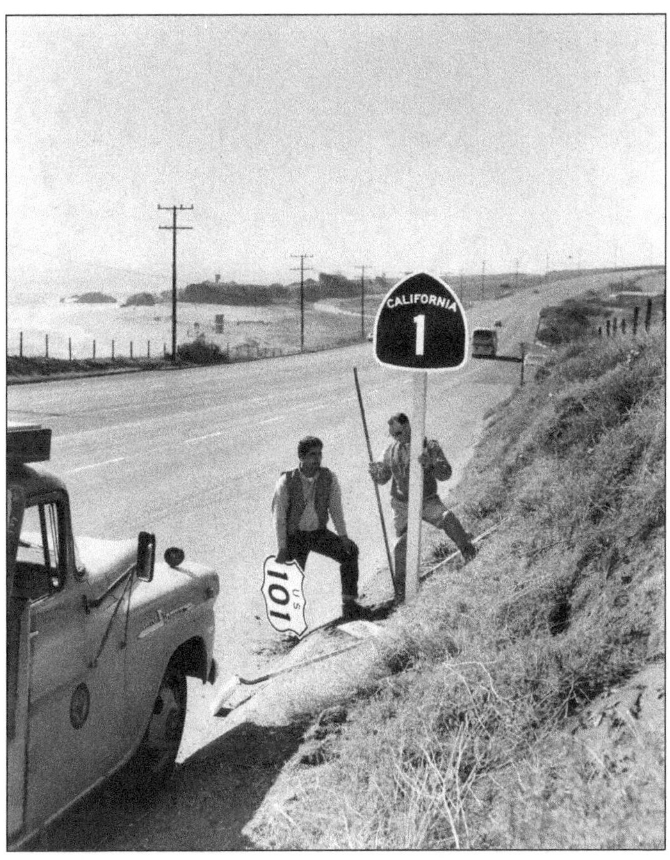

Two state engineers replace a US Route 101 road sign with a new California Highway 1 sign at Point Mugu. The photograph was featured on the cover of *California Highways* magazine in the fall of 1964, and in the same year, the coast highway in the counties of Orange, Los Angeles, Ventura, Santa Barbara, and San Francisco, to its northern terminus near Liggett, were designated as Highway 1. The installation of the Highway 1 signs began after California began numbering its highways in 1934. The coast road had several names through the years, including Route 56, Route 60, Highway 3, Route 3, and US Route 101 Alternate. (Copyright California Department of Transportation. All rights reserved.)

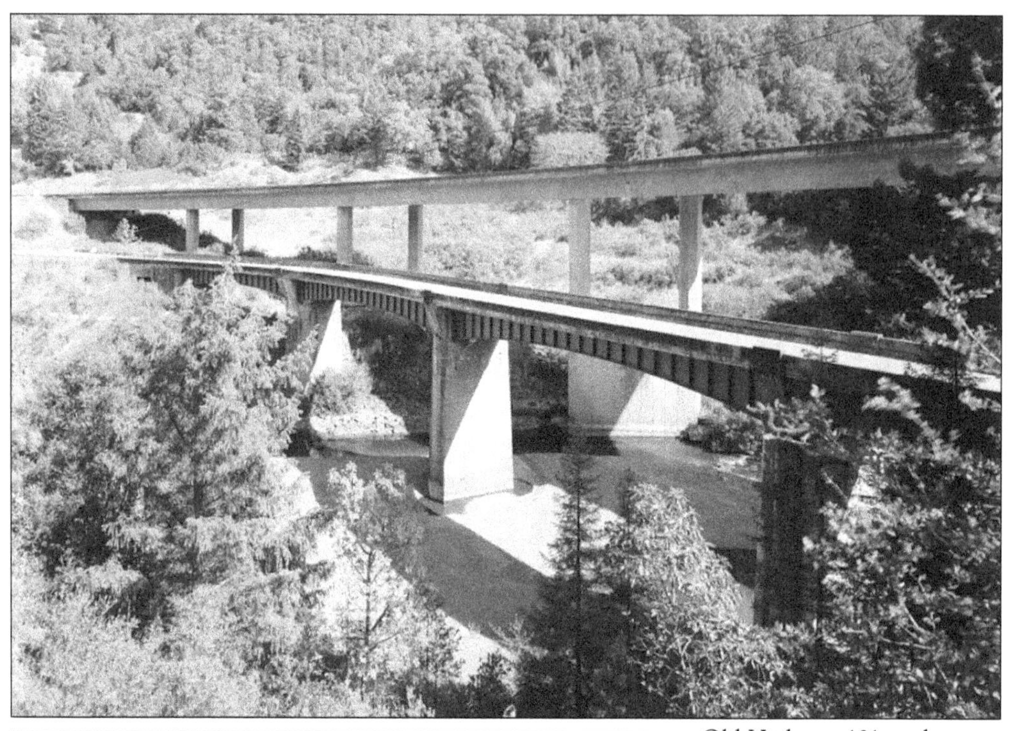

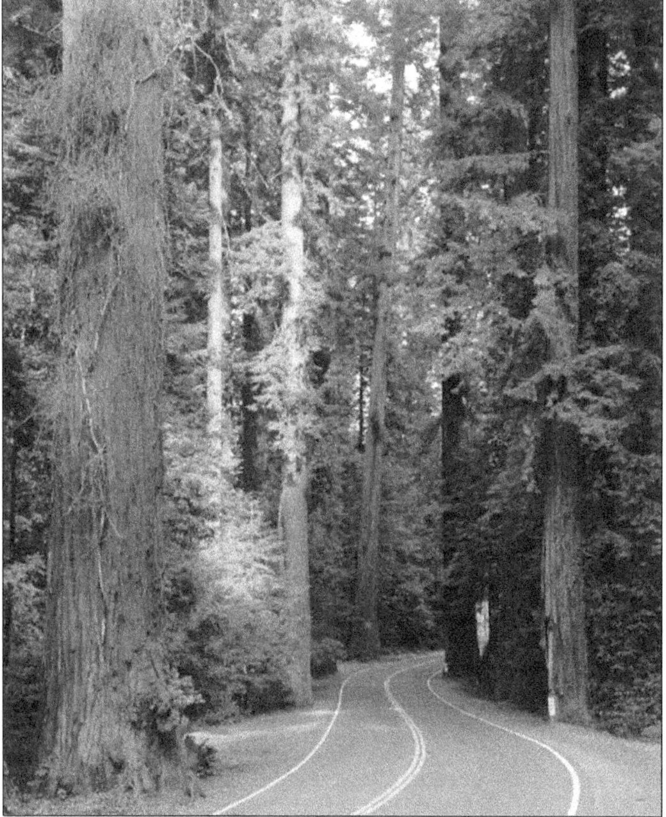

Old Highway 101 at the Smith Point Bridge in Humboldt County is seen here in 1968. Also called the Redwood Highway, it is one of the most scenic segments of the highway. The new highway was constructed higher than—and parallel to—the old road. The bridges are located in Redwood National Park.

Avenue of the Giants is probably the most scenic segment of the highway in Northern California, as seen here in 1980. The new highway, State Route 254, is the previous alignment of Highway 101. The two-lane road is shadowed by coastal redwoods and follows alongside the Eel River.

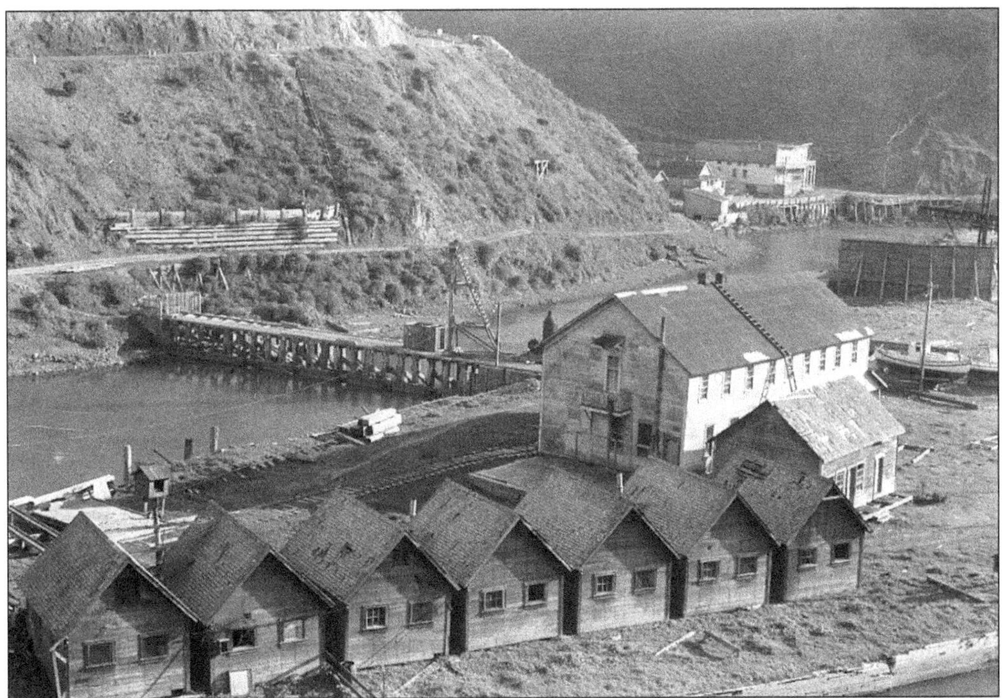

Pictured is an abandoned lumber community in Humboldt County. Housing was provided for workers and their families. Small houses and the mill are seen here in 1938. The lumber industry in Northern California boomed in the 1850s and supplied lumber to the growing city of San Francisco. By 1854, there were nine mills along the bay. Today, two companies established in the mid-1850s still remain. Below is another view of the lumber community showing a log dam, which was used to prevent or reduce sediment runoff into streams. The dam decreased water velocity and allowed sand, silt, and mud to settle out behind check dams.

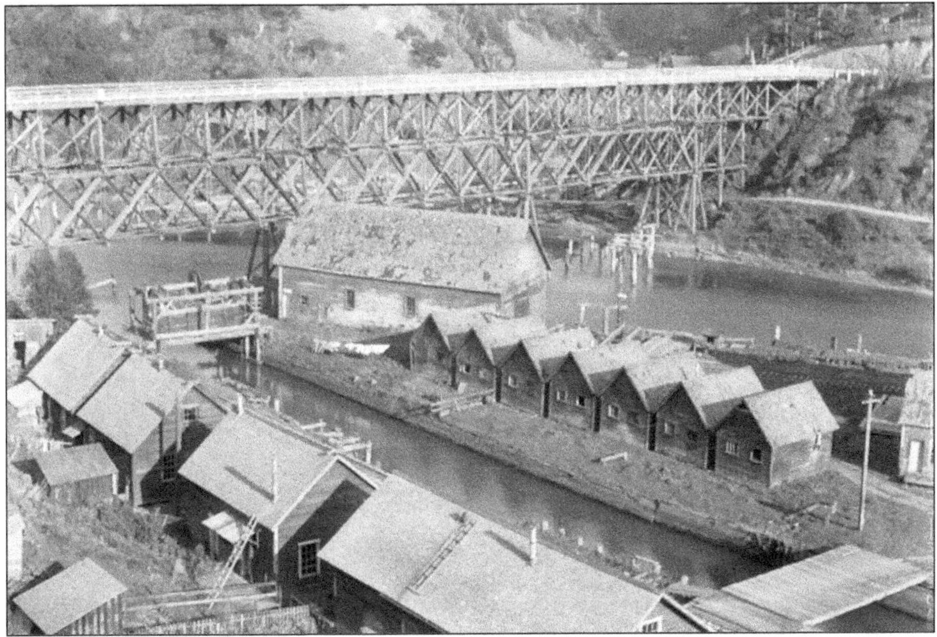

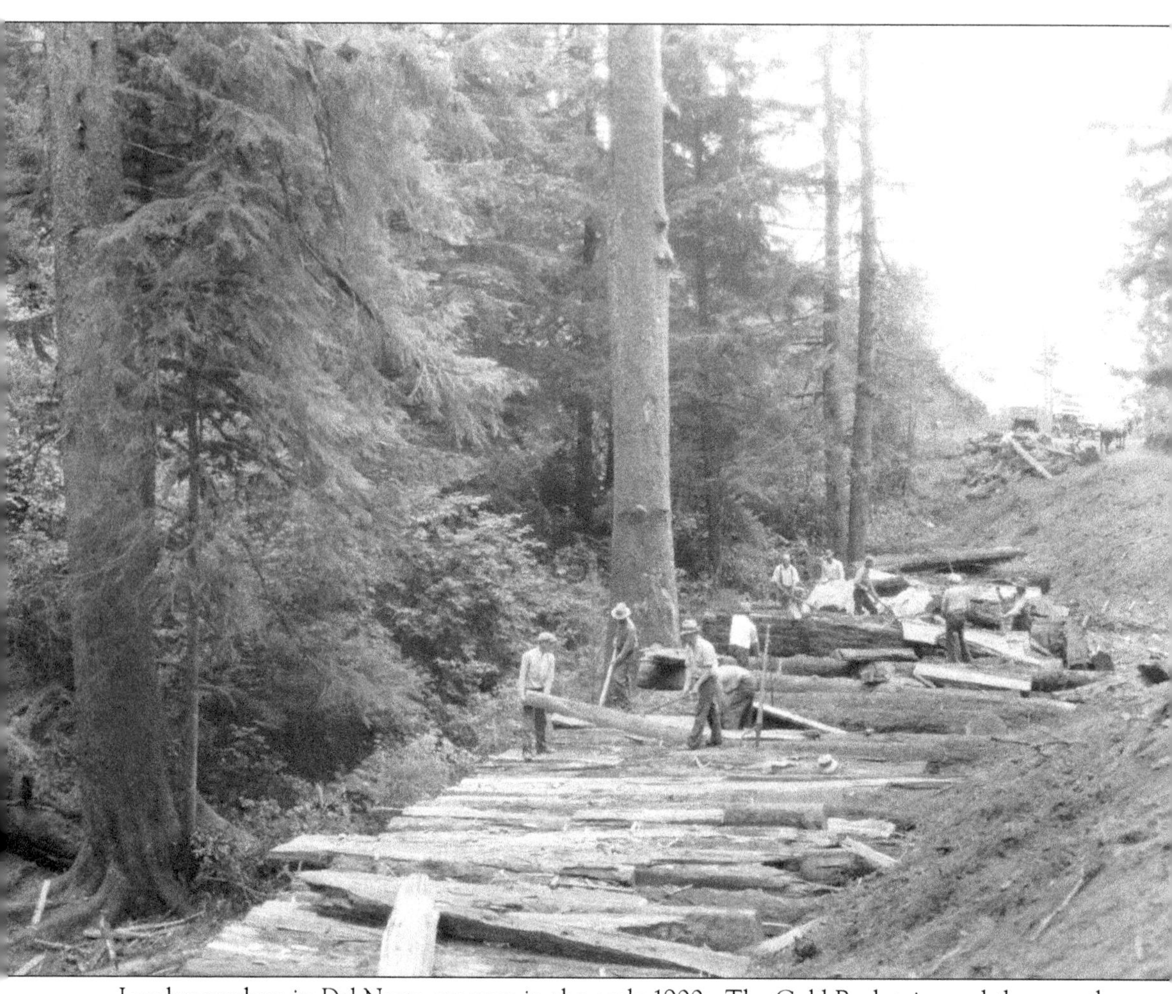

Lumber workers in Del Norte are seen in the early 1900s. The Gold Rush triggered the growth of the lumber industry in California. San Francisco's population boomed from 2,000 in 1849 to approximately 59,000 in 1855, resulting in a high demand for lumber.

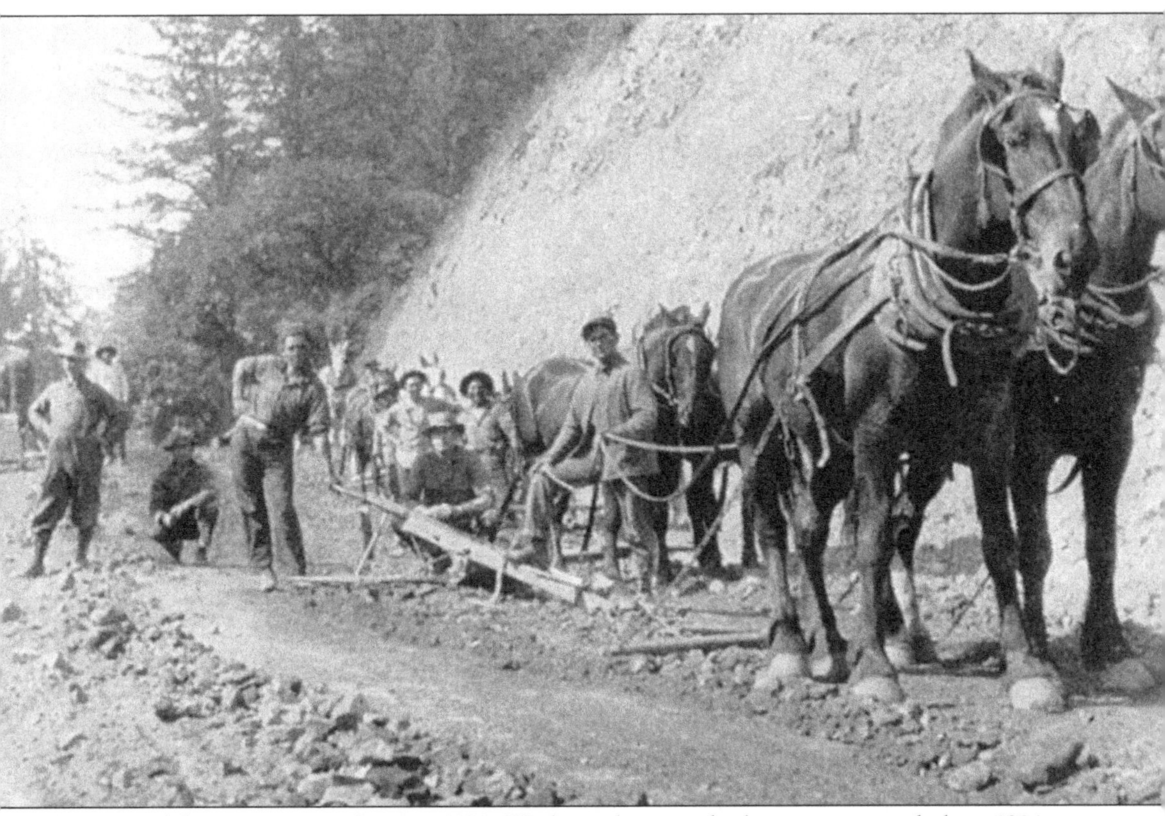

A convict labor crew is seen here in 1928. Work on the coast highway came to a halt in 1924 due to funding issues and a shortage of laborers. When work resumed in 1928, convict labor helped solve the problem of finding enough people willing to work in remote regions of the state.

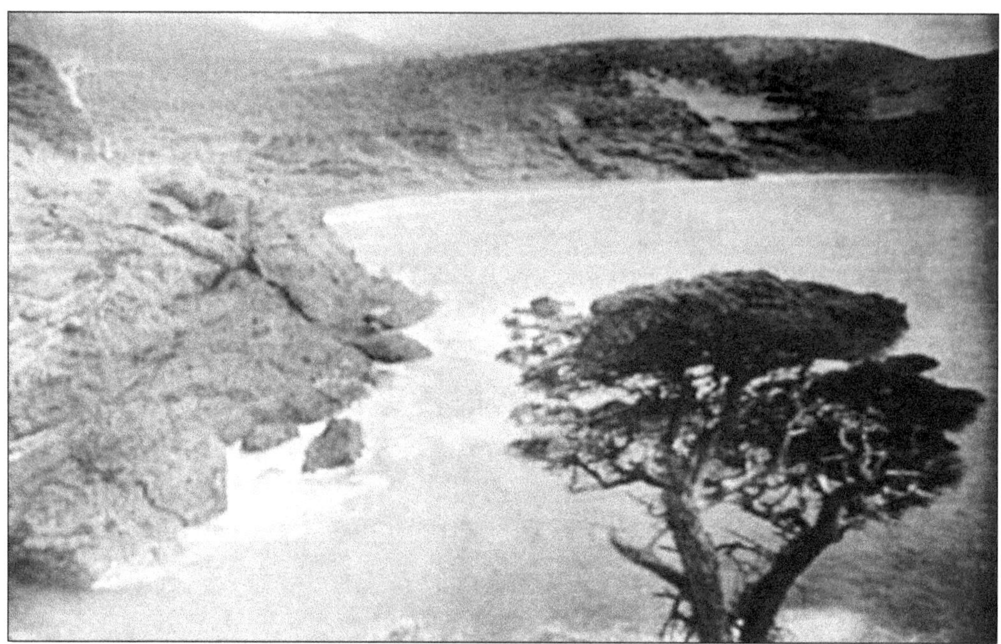

The city of Carmel, also called Carmel-by-the-Sea, is located along the Monterey coast. Founded in 1902, it was one of the most isolated areas before the coast road was constructed. Founded in 1770, Mission San Borromeo de Carmelo was located in a nearby settlement but relocated by Franciscan monk Junípero Serra to Carmel.

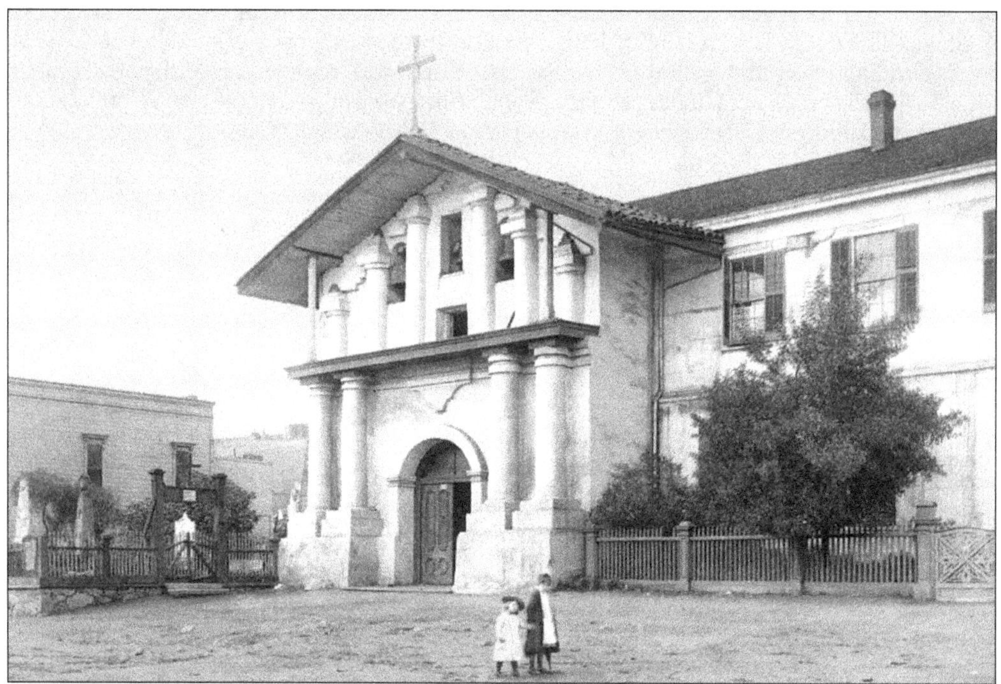

Founded in 1776, Mission Dolores is the oldest building in San Francisco that remains intact. It is formally named Mission San Francisco de Asís and is one of the 21 California missions built along El Camino Real, "The Royal Road," which the California coast route follows. (Mission Dolores.)

The coast highway at Waddell Creek Bridge is seen here. The bridge was built in 1947. Located in Santa Cruz, it is named for both the creek and the watershed that empties into the Pacific Ocean at Waddell Beach. Waddell Bluffs (pictured below)—named after William Waddell, who was in the lumber industry in the 1860s—are along Highway 1 and stretch 1.4 miles along the coastline. The vertical bluffs are comprised of Santa Cruz mudstone. Although steep and prone to erosion, they are still standing today. Before construction of the highway, stagecoaches and horse carriages were able to pass the bluffs only during low tide. (Below, Robert C. Pavlik.)

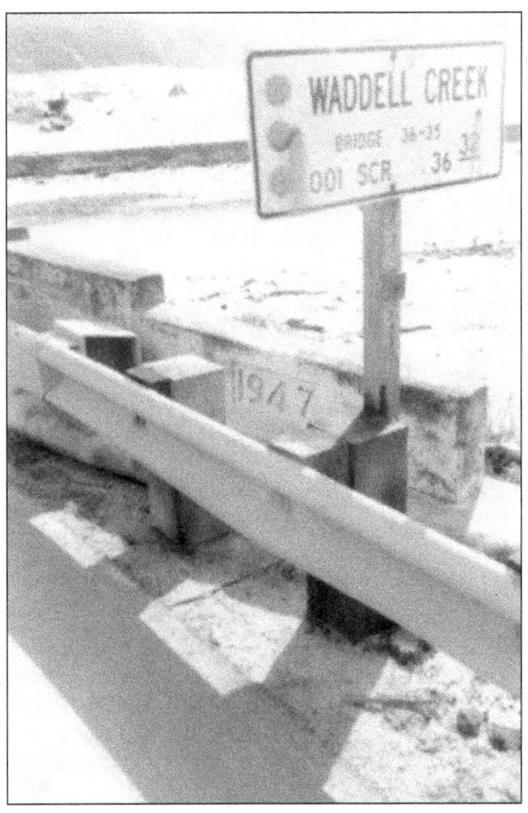

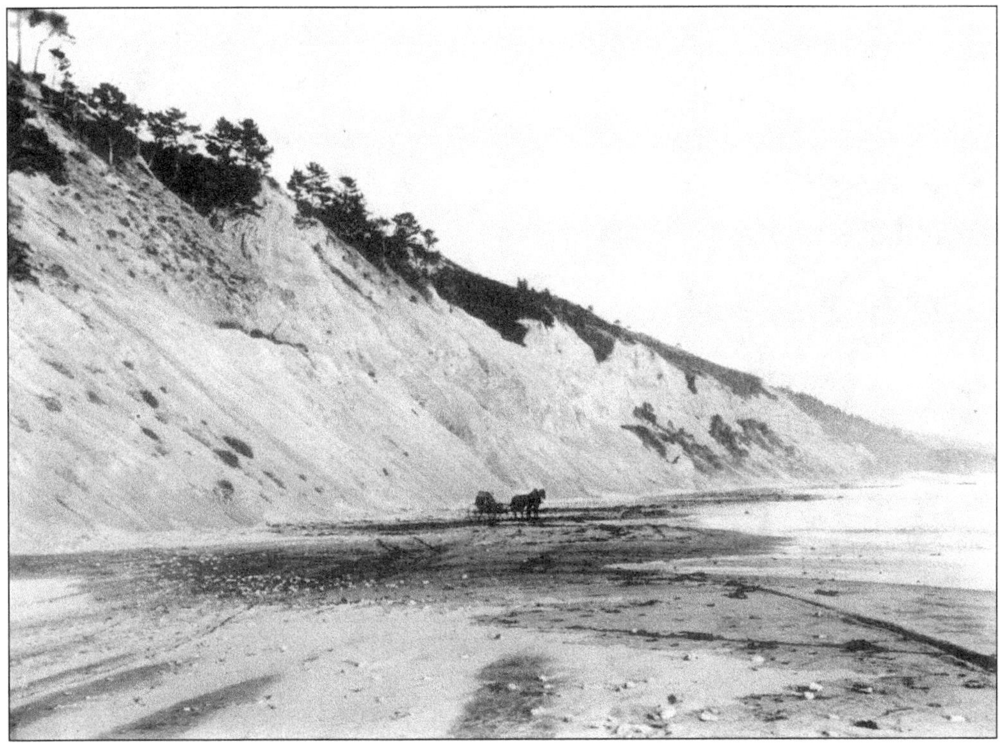

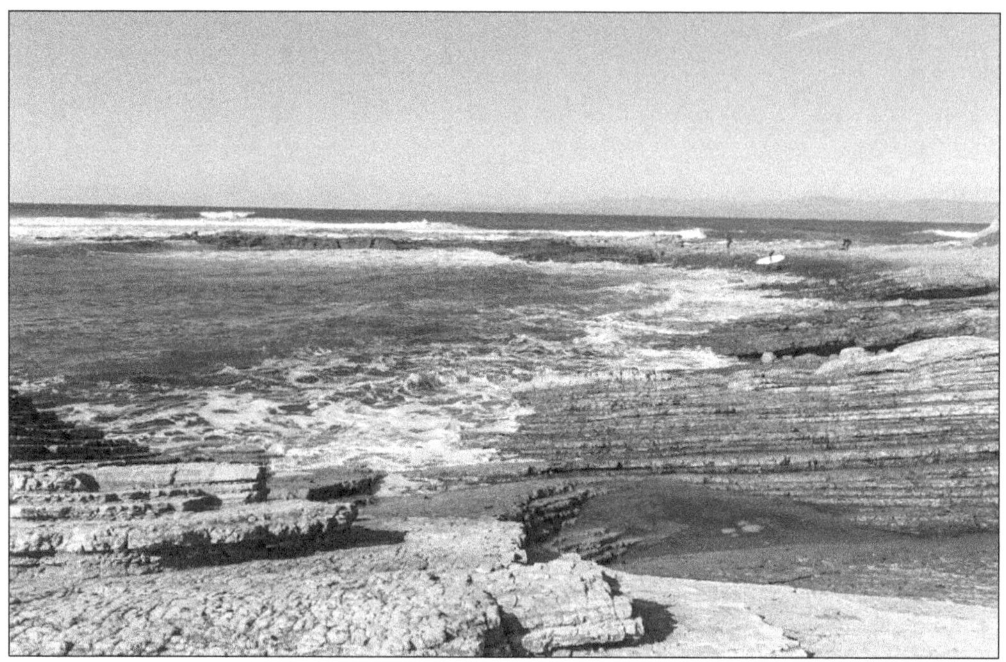

Pictured is a section of the Morro Bay coast in the heart of the central coast. The Spanish Portola expedition camped near Morro Bay in 1769. Franciscan missionary Juan Crespi, who was part of the expedition, noted in his diary, "We saw a great rock in the form of a round morro."

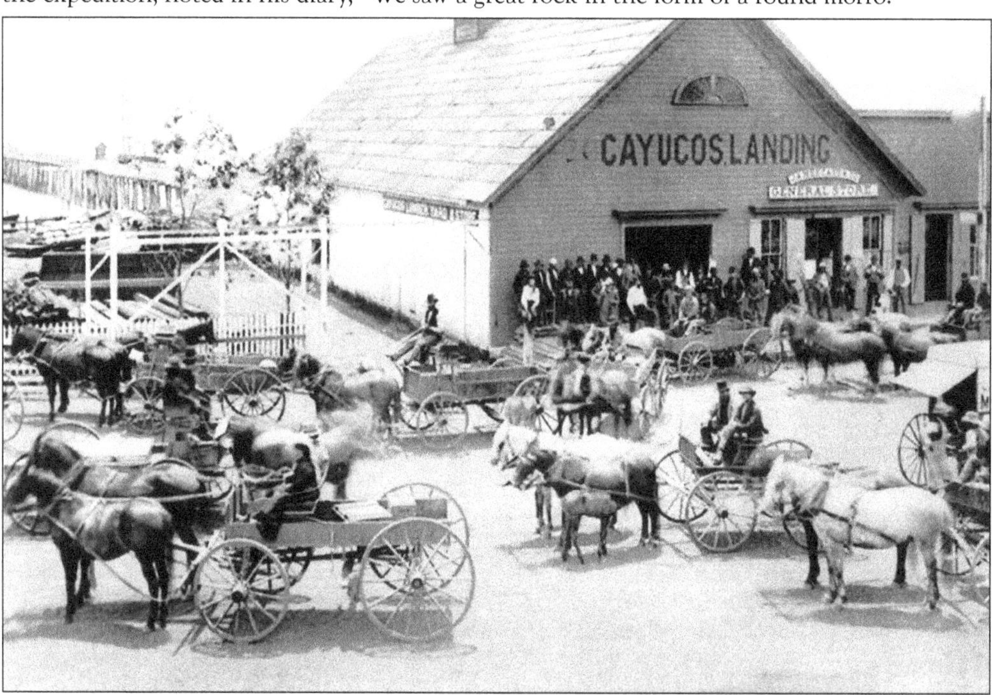

Cayucos Landing, lumber yard, and general store are shown here in the mid-1800s. The California Gold Rush produced a high demand for lumber as populations and communities grew. Shipping lumber along the coast was difficult. Ship landings came into existence along the coast at small coves, also called "dog holes." (Boeker Street Trading Company.)

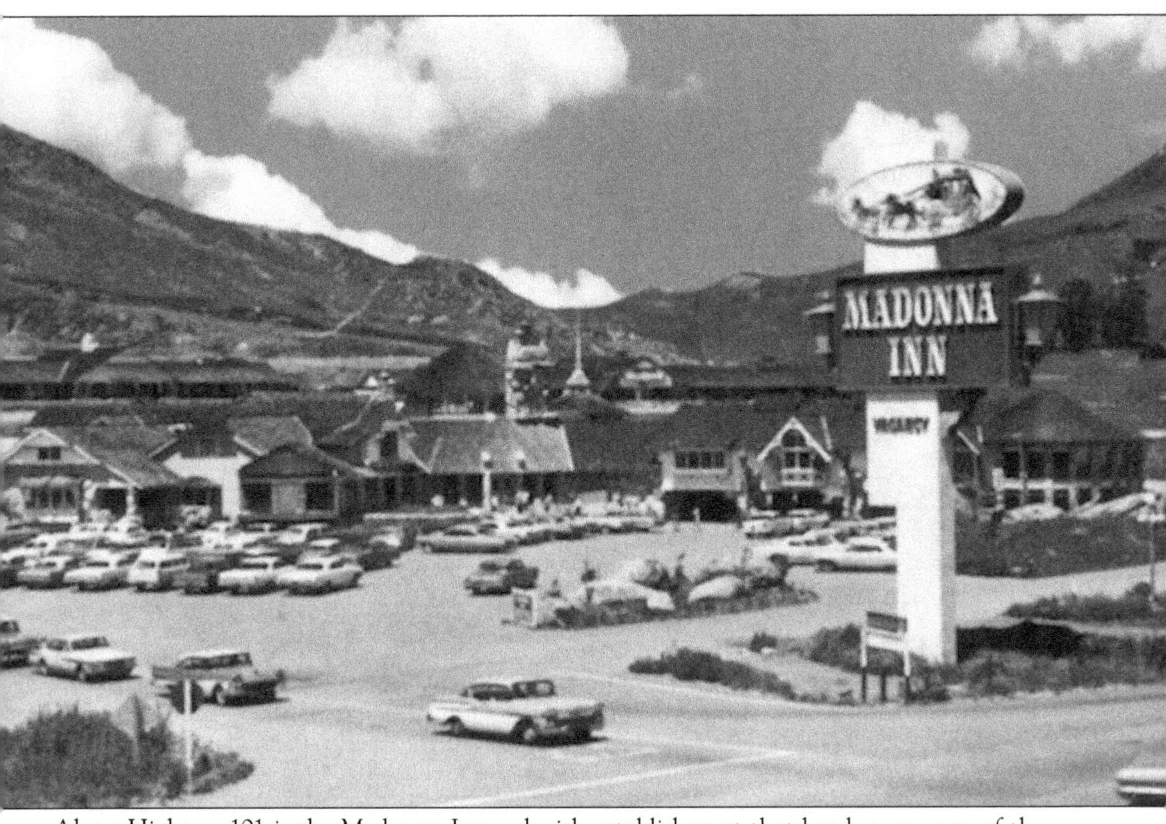

Along Highway 101 is the Madonna Inn, a lavish establishment that has become one of the famous landmarks of the central coast. It began with 12 rooms in 1958 and has grown to 110 unique themed rooms. The property is over 1,000 acres. The inn is pictured here in the 1960s. Its interior décor is filled with leaded glasswork and etched glass windows. The leaded glass inserts by the fireplace symbolize construction, lumber, and cattle, which are Madonna's enterprises. The hand-carved balustrade in the dining room is from William Randolph Hearst's San Simeon warehouse. Wood carvers were brought in to hand carve doors, beams, railings, and other adornments. (Boeker Street Trading Company.)

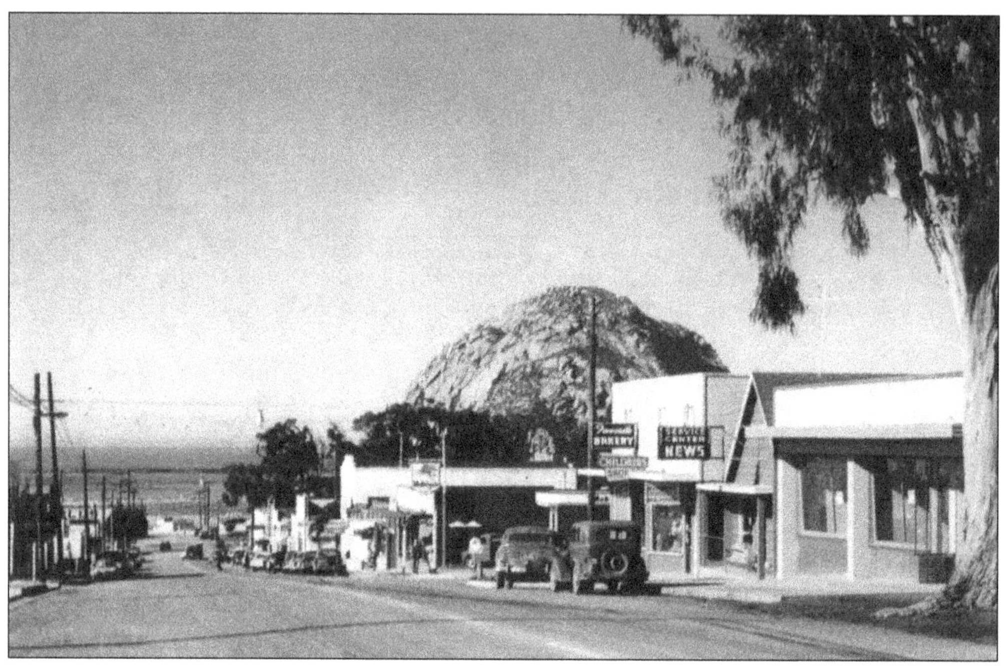

The small town of Morro Bay is seen here in the 1930s. It was founded by Franklin Riley in 1870 primarily as a port for exporting dairy and ranch products. Riley also helped build the town's wharf, which later became the Embarcadero. Morro Rock in the background is a famous landmark of the central coast. (Boeker Street Trading Company.)

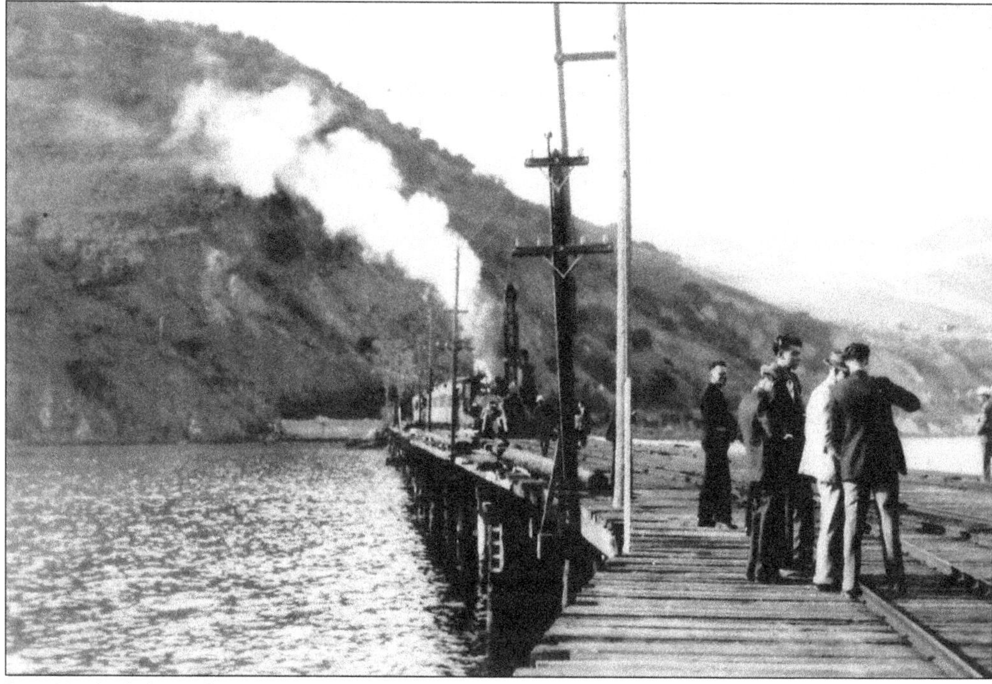

Port Harford along the San Luis Obispo bay is seen in 1940. The port was established in 1871 after a pier and horse-drawn railroad was built by John Harford. The Pacific Coast Railway once ran through the port on the San Luis Obispo to Santa Maria route. (Boeker Street Trading Company.)

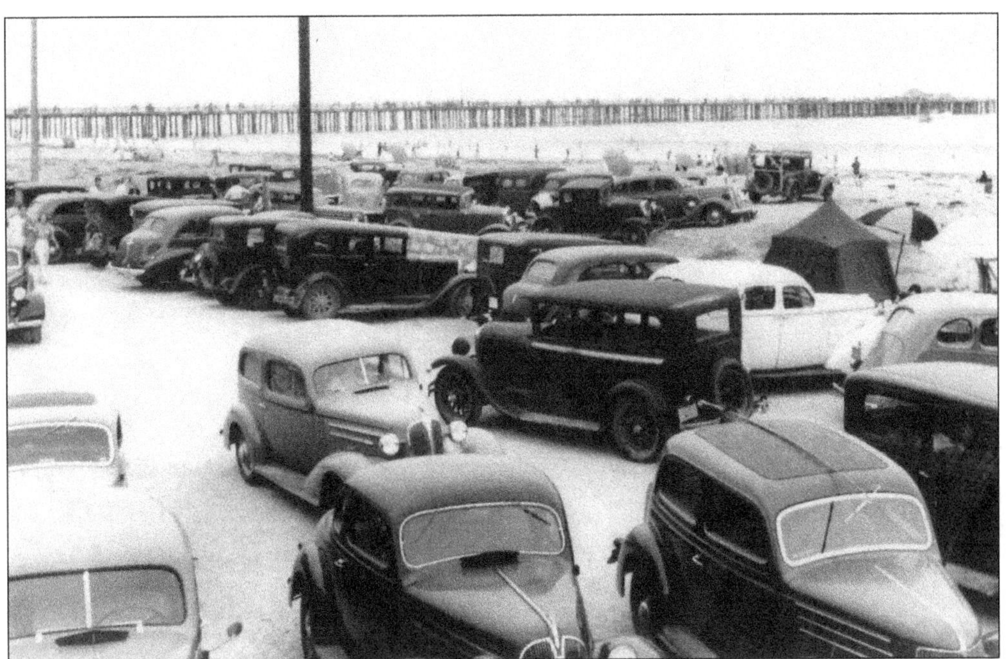

Automobiles park on the sand at Avila Beach in 1930. Established in the late 19th century and named after Miguel Avila, one of the grantees of the former Rancho San Miguelito in 1842, the town of Avila Beach was at one time the main shipping port for San Luis Obispo. (Boeker Street Trading Company.)

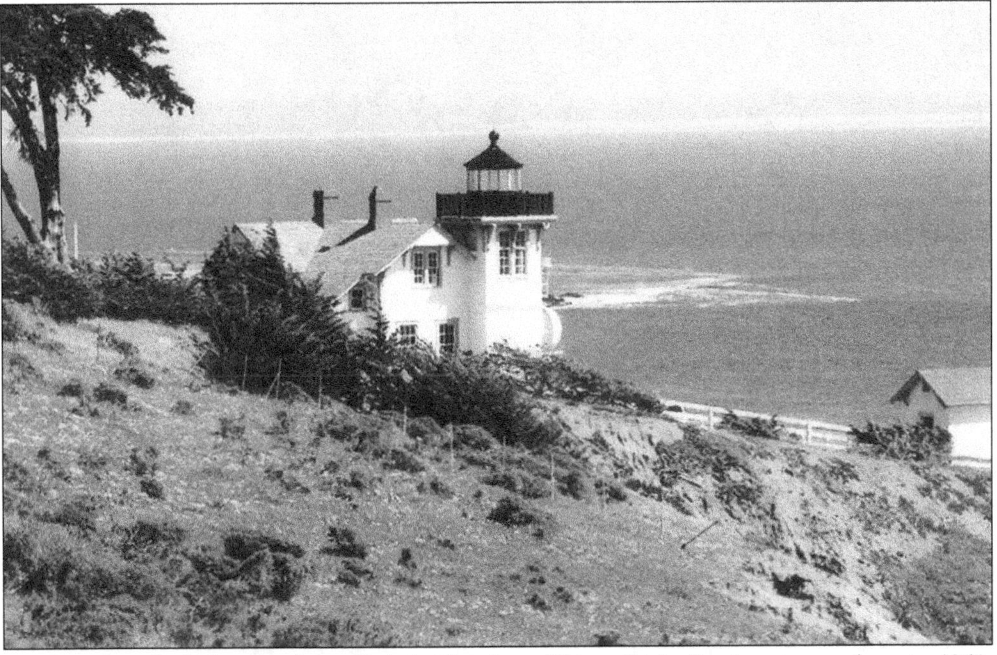

Port San Luis Lighthouse, also called the San Luis Obispo Light Station, is seen here in 1960. Founded in 1890, it has been the San Luis Bay beacon for over a century. The lighthouse served nearby Port Harford when the port was booming and averaged approximately 400 ships per year. (Boeker Street Trading Company.)

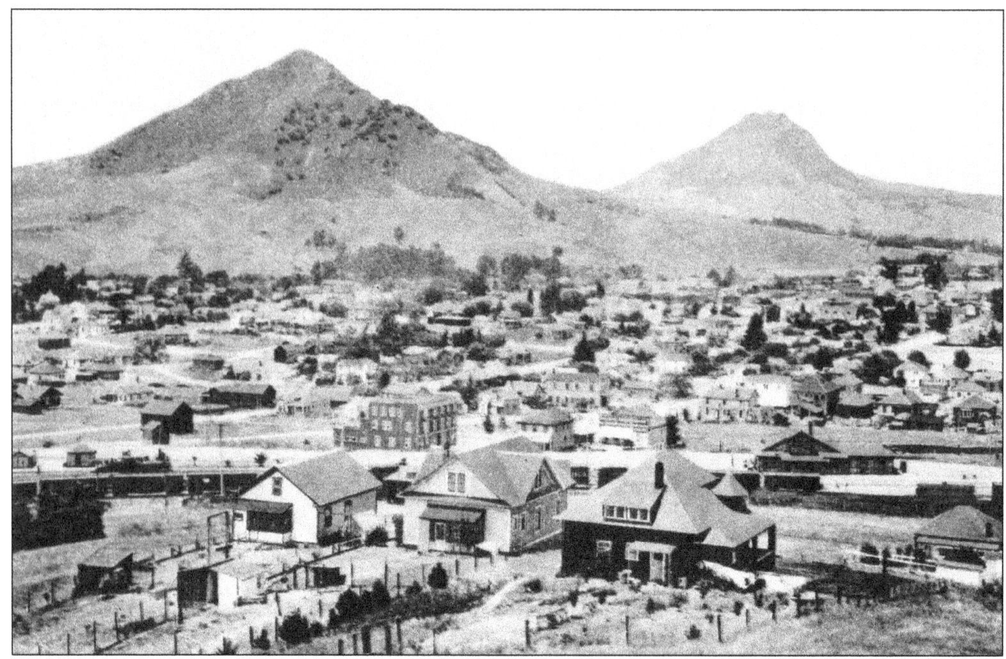

This is an early 1900s view of San Luis Obispo. The city is one of the oldest communities in California. Founded in 1772 by Franciscan monk Junípero Serra of the Portola expedition of 1769, Mission San Luis Obispo de Tolosa was built in 1772, and became the town of San Luis Obispo. (Boeker Street Trading Company.)

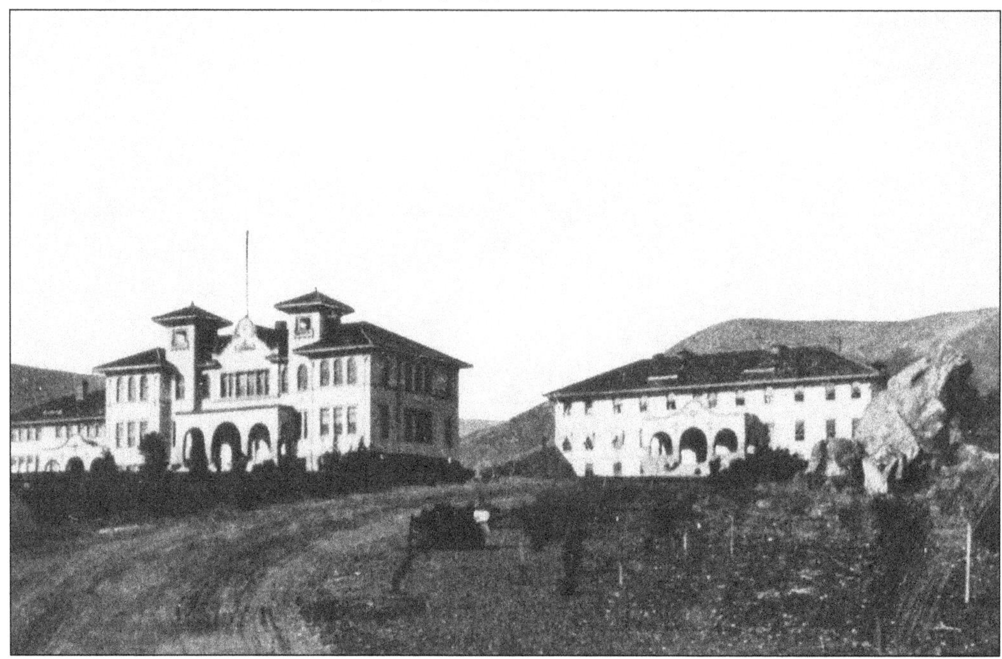

California Polytechnic State University in San Luis Obispo is pictured in the 1930s. Founded by Myron Angel in 1901 as a vocational high school, it later became a college in 1940 and then a university in 1972. It is one of two polytechnic universities in the California State University system. (Boeker Street Trading Company.)

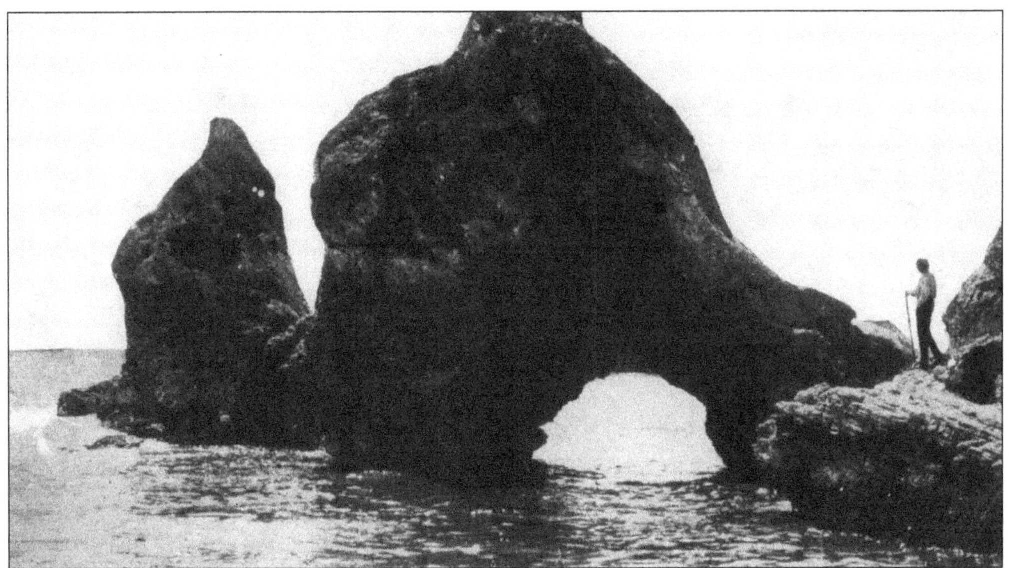

Arch Rock at Shell Beach is seen here in the 1960s. Ocean arch rocks are formed where limestone sea stacks have eroded over time. The softer part of the rock erodes away, forming an alcove. Further erosion of the alcove results in an arch. (Boeker Street Trading Company.)

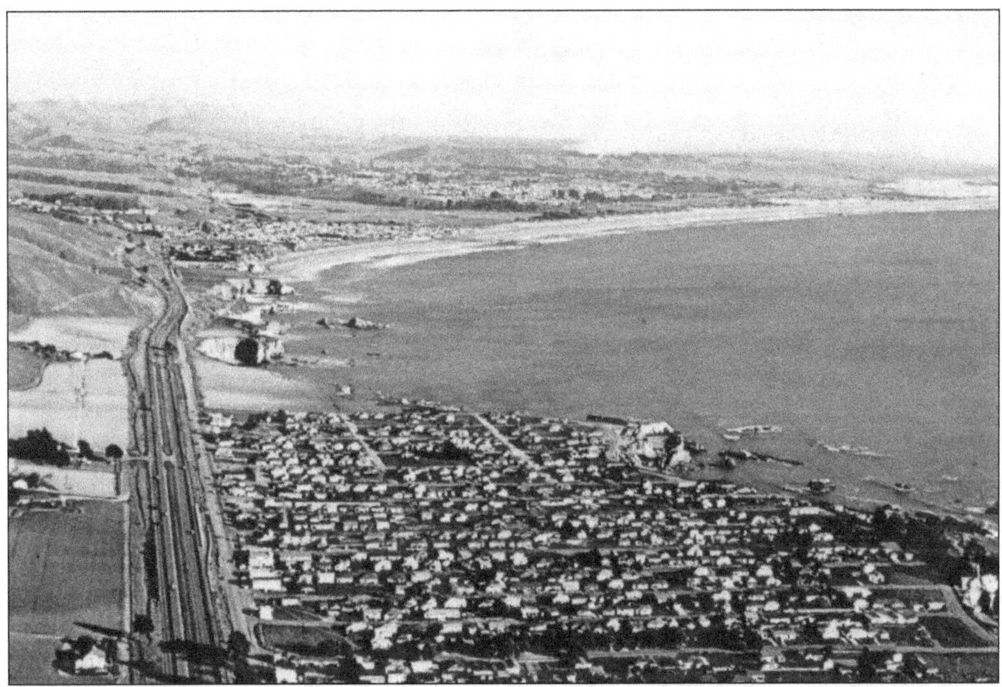

This 1950 aerial view of Pismo Beach includes Shell Beach. Shell Beach is a neighborhood in Pismo Beach, which is known for its nine separate beaches just below its cliffs. It is highly prized for its picturesque coves and bluff tops, but the sand sections of the beaches shrink during high tide. (Boeker Street Trading Company.)

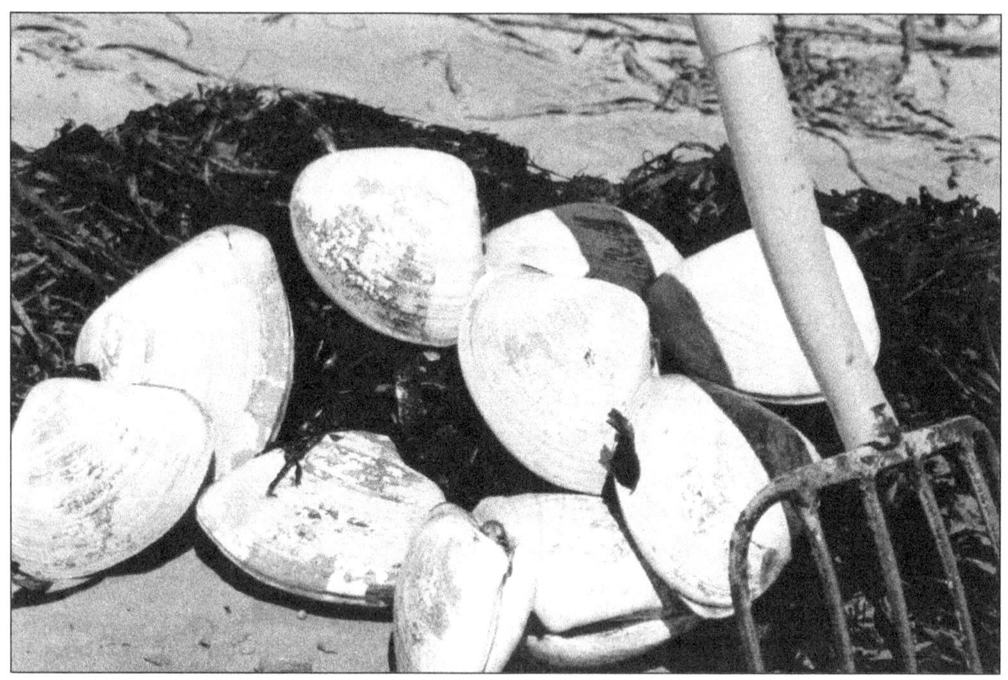

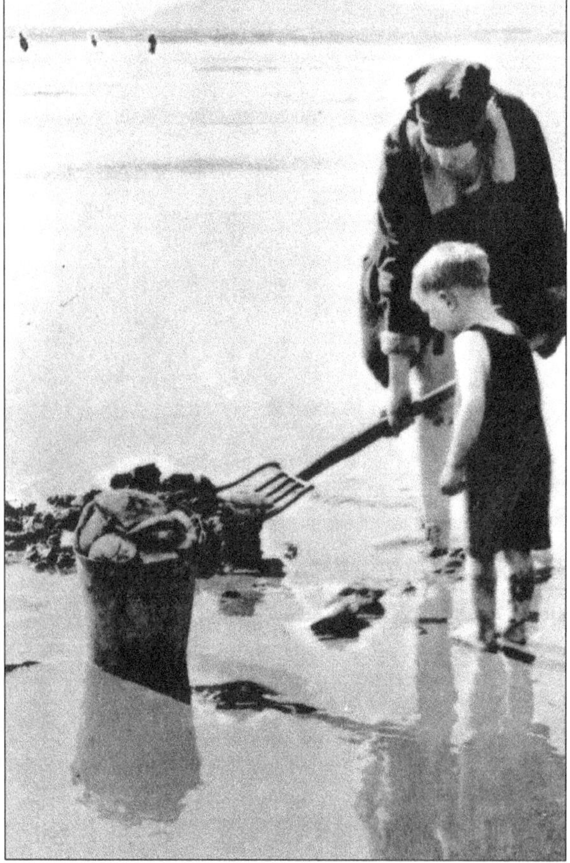

Pismo Beach's famous Pismo clams are seen in the 1950s. Pismo clams measure four and a half to eight inches in diameter. One of the largest species of clam, they are found only at Pismo. Once free for the taking, a clam digger could take away a hundred or more clams a day. (Boeker Street Trading Company.)

This 1950s photograph shows a woman digging for clams while a little boy watches. Her bucket is filled with large clams. Clam digging in ankle-deep water was a popular activity at Pismo up until around 1981, when clams began to vanish as a result of over-clamming and sea otter foraging. (Boeker Street Trading Company.)

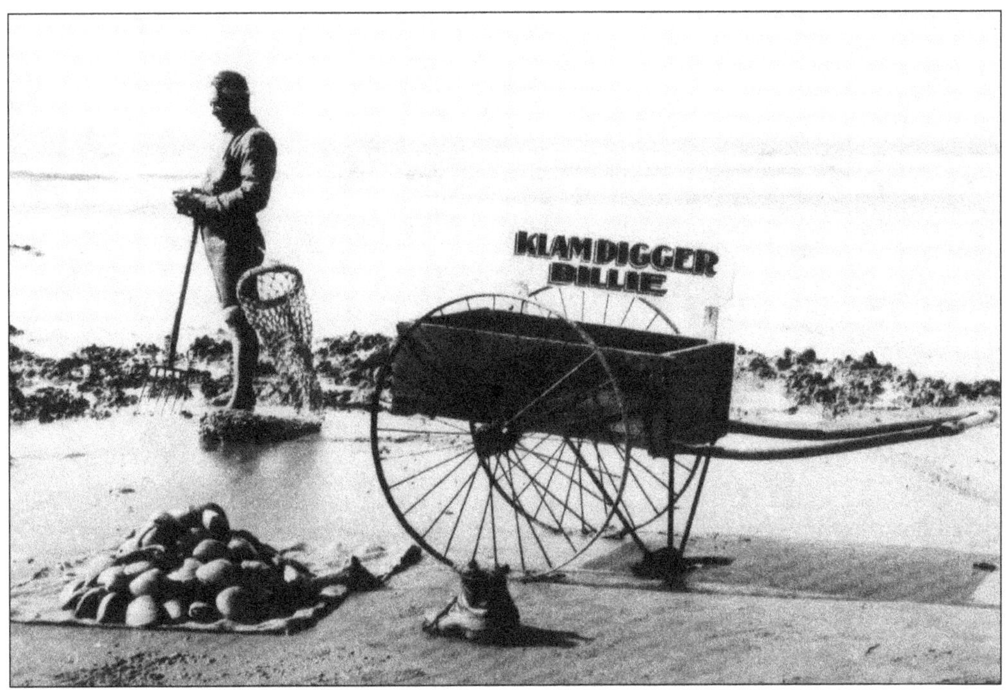

This 1904 photograph shows "Klamdigger Billie" at Pismo Beach with his cart of clams for sale to those who preferred not to get their feet wet. A large pile of clams can be seen resting on burlap below his cart. Today, a fishing license is required to take clams of at least four and a half inches. (Boeker Street Trading Company.)

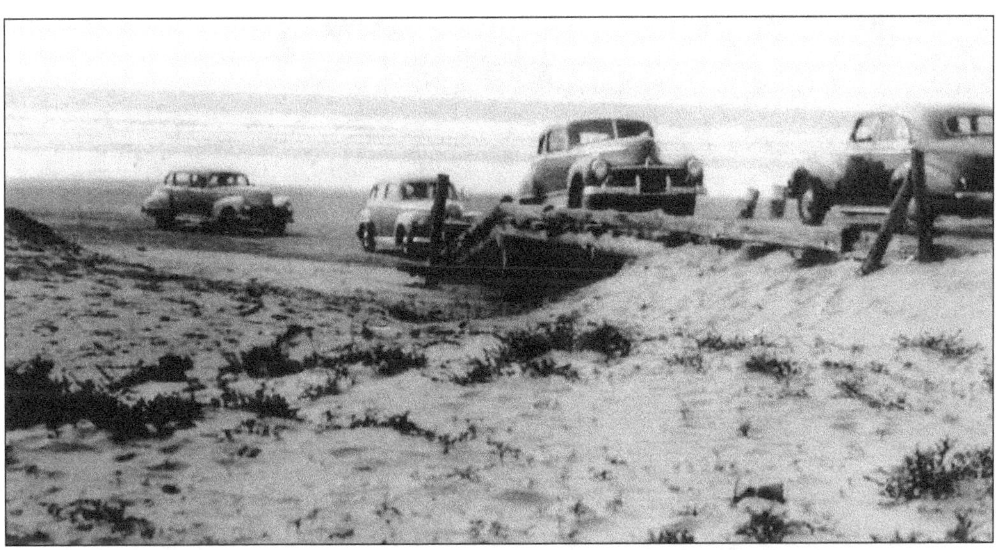

Automobiles are seen on Oceano's beach in the 1940s on the clam approach. The motorists were likely clam diggers because this section of the beach was known as a clam reserve. Oceano's beach is also part of neighboring Pismo Beach, where people also go to dig for clams. (Boeker Street Trading Company.)

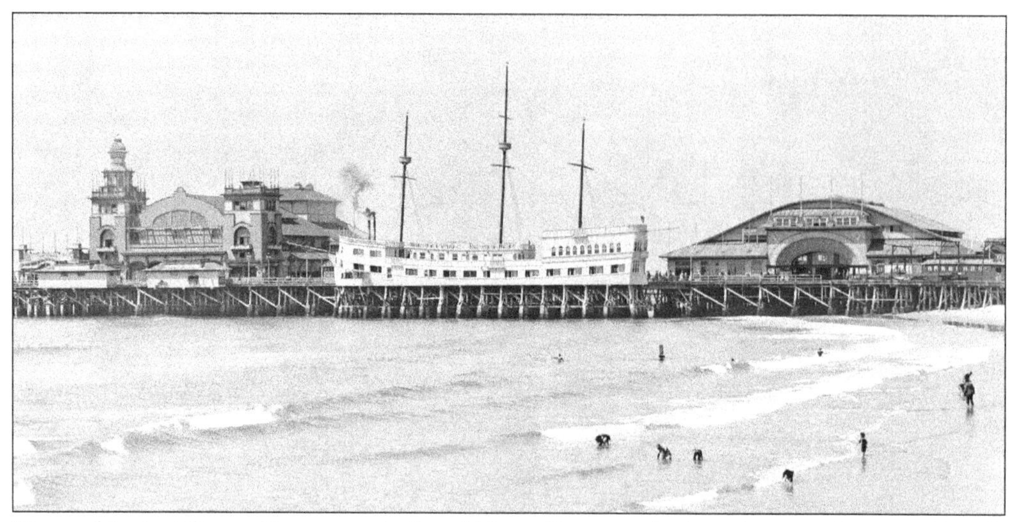

Pictured is the Kinney Pier, named after Abbott Kinney, who bought several acres of beach property that is now present-day Venice Beach. The 1,200-foot pleasure pier was built in the 1900s, and all of the structures on the pier were of Venetian style. (Jon B. Lovelace Collection of California Photographs.)

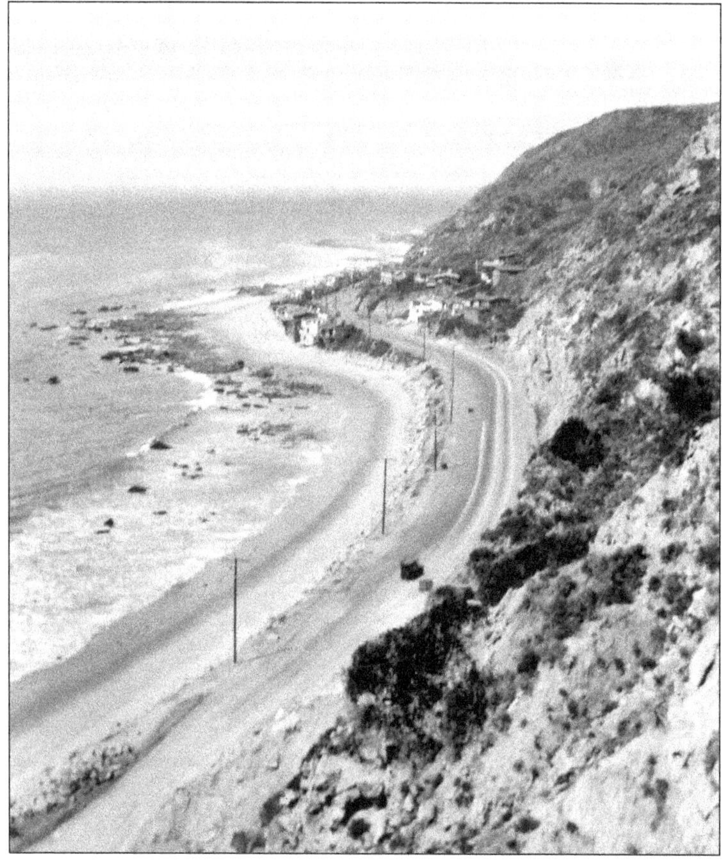

The Pacific Coast Highway is seen in 1933 along the Santa Monica coast curving around the Santa Monica Mountains. It was a two-lane road that ran along unprotected mountains. The mountainside would later require extensive slope excavations. (Copyright 1933 California Department of Transportation. All rights reserved.)

The Pacific Coast Highway along the Santa Monica coast is seen here after hillside reinforcement. Reinforcement measures included slope ditches, installation of riprap, use of laminated timber guardrails on timber posts, and installation of underdrains. (Copyright California Department of Transportation. All rights reserved.)

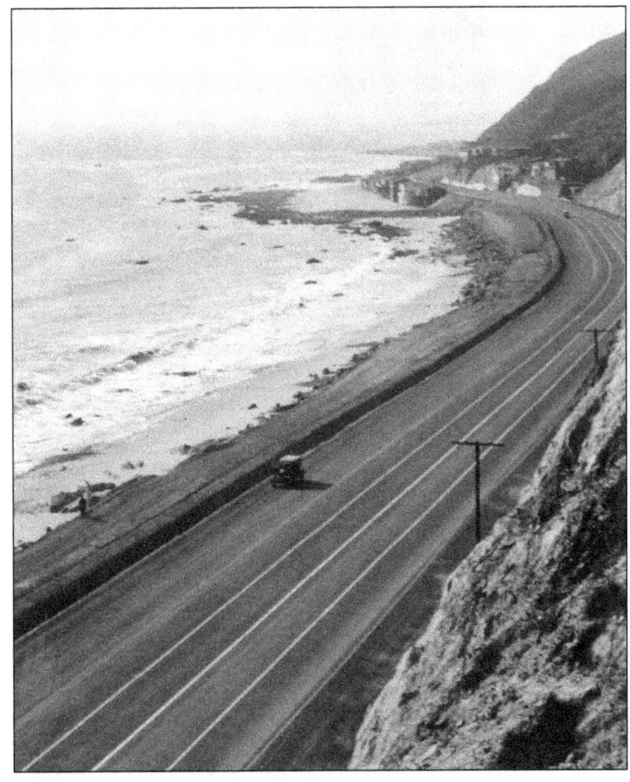

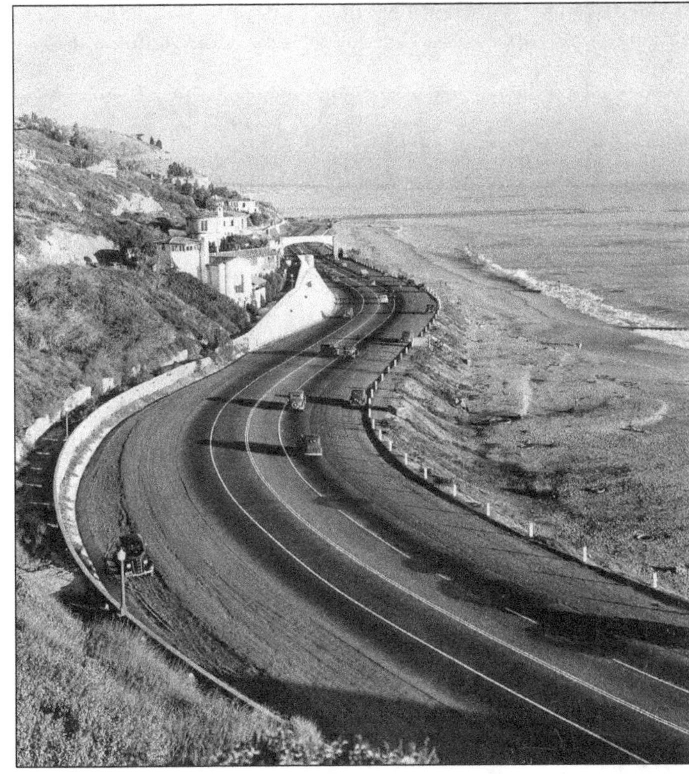

The Pacific Coast Highway, formerly called Roosevelt Highway, is seen looking south along the Santa Monica coast near Pacific Palisades in 1938. The hillside slopes along this section underwent a reinforcement project that aimed to reinforce the unstable hillsides along the Santa Monica coast. (Copyright California Department of Transportation. All rights reserved.)

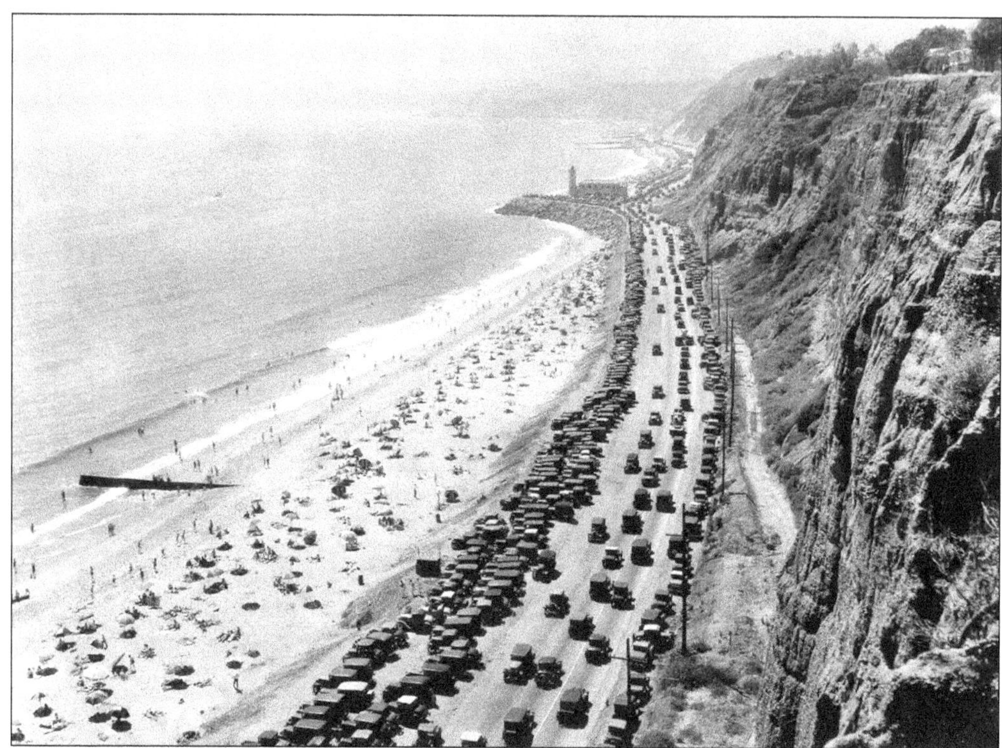

Looming bluffs along the Santa Monica coast are pictured in 1930. Significant erosion can be seen along the bluffs that span the distance from Santa Monica to beyond Pacific Palisades. The beachside that adjoins the highway also appears weak from erosion. (Copyright California Department of Transportation. All rights reserved.)

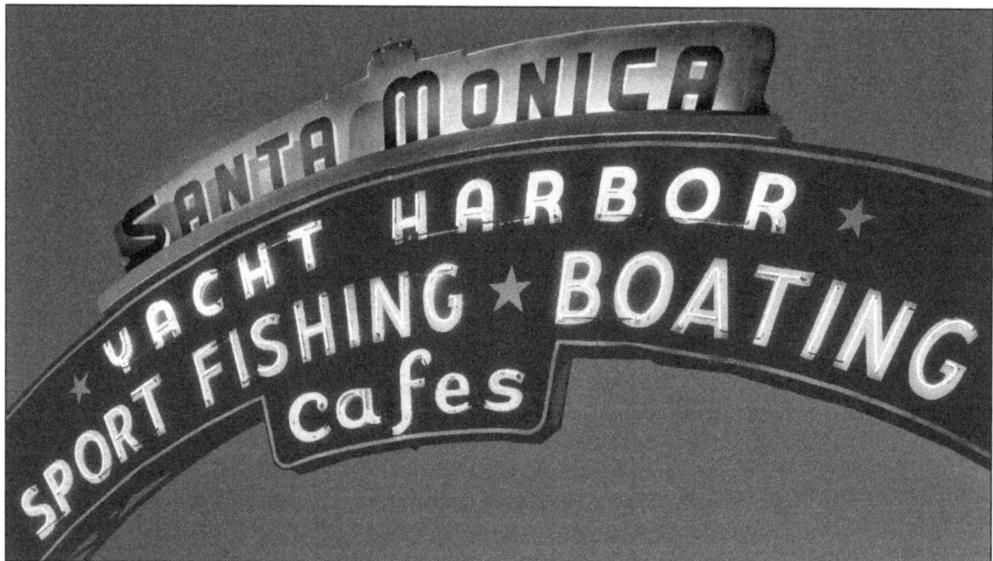

The Santa Monica Yacht Harbor neon sign, installed in 1940, is one of the most photographed pier signs on the Southern California coast. The harbor once aimed to become Southern California's premier boat harbor, but its breakwater began to sink in the 1960s, causing water traffic to significantly decline. (Library of Congress.)

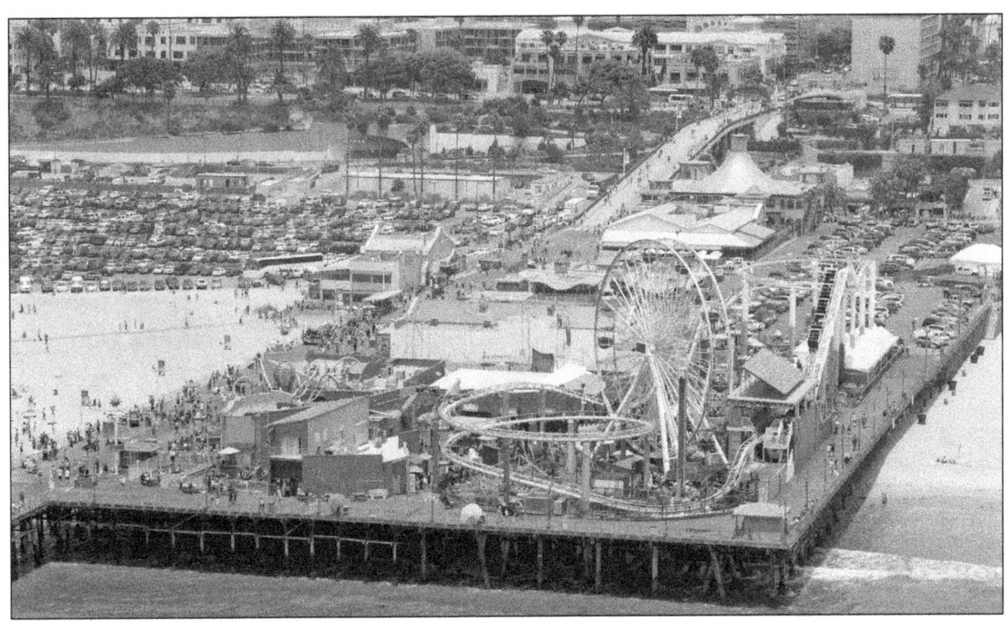

Santa Monica Pier has been one of Southern California's most visited attractions along the coast. Pacific Coast Highway traffic by the pier has always been heavy and remains congested today. Built as a pleasure pier in 1916, another name for an amusement park, it is still in operation and continues to be a popular destination. It is a double-joined pier, comprising two separate piers, as shown below. The first pier was built long and narrow in 1909 and transported sewage directly into the ocean through pipes underneath. The second pier was wider and was the site for the pleasure pier. (Both, copyright California Department of Transportation. All rights reserved.)

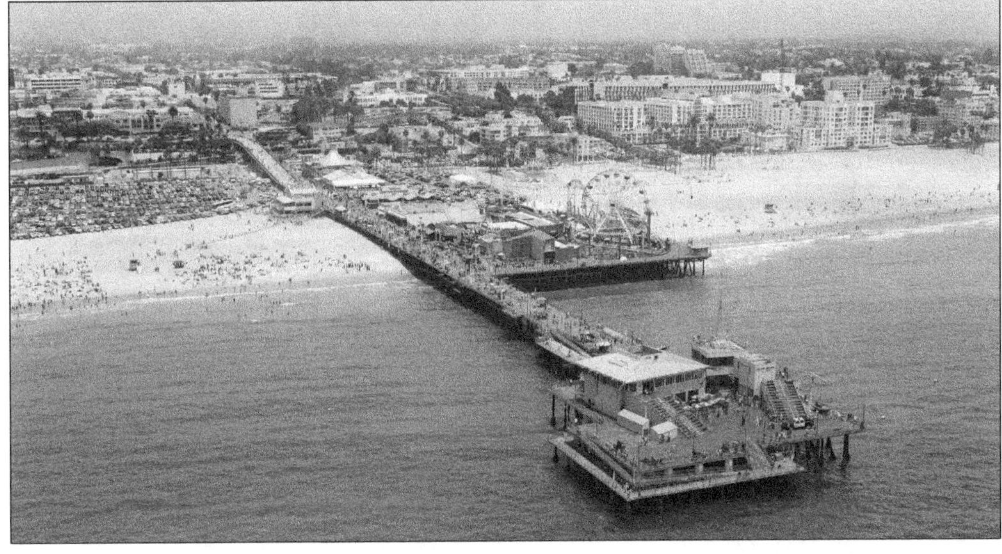

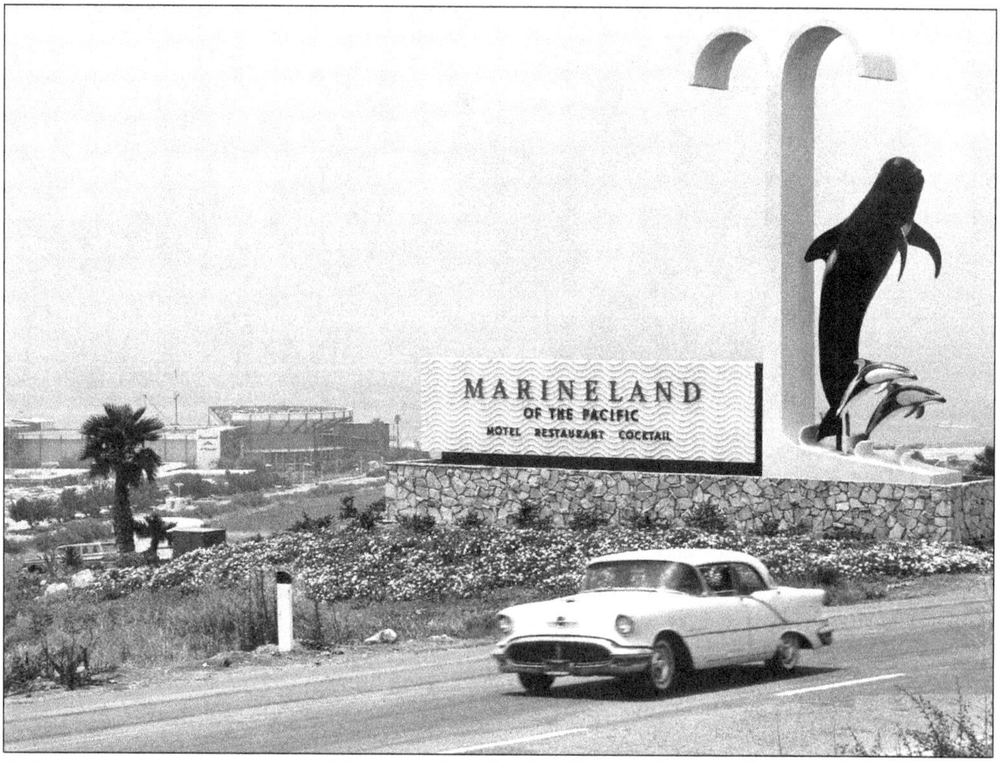

A favorite coast attraction was Marineland of the Pacific, which opened in 1954 and featured a large population of marine life, including killer whales. The oceanarium occupied 90 acres and was in operation from 1954 to 1987. Sea World San Diego purchased the oceanarium. (Jon B. Lovelace Collection of California Photographs.)

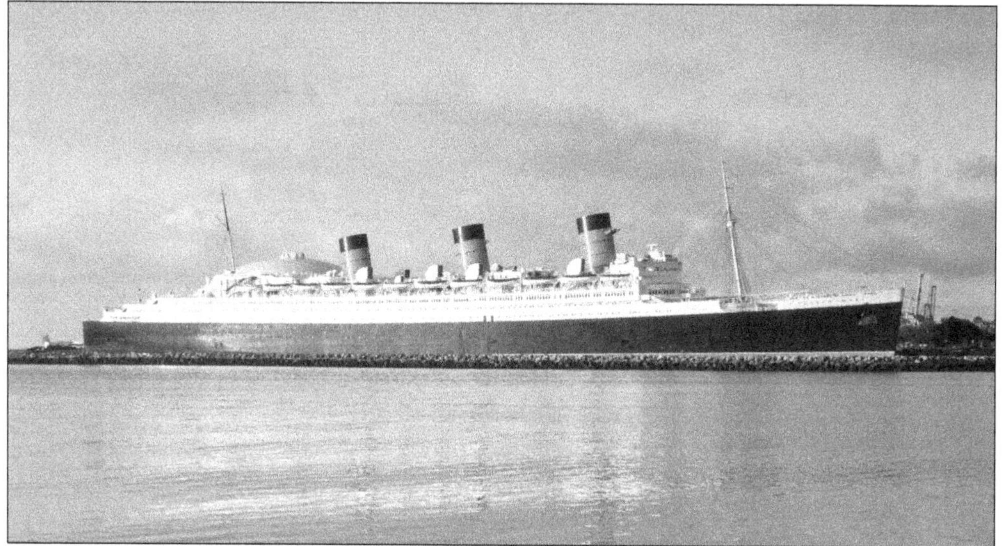

Pictured is the retired RMS *Queen Mary*, an ocean liner that sailed the North Atlantic from 1936 to 1967. It rests in the Port of Long Beach. Its first voyage was in 1936; it was also used as a troop ship ferrying Allied soldiers during World War II. After the war, it was returned to passenger service until its retirement in 1967. (California Department of Transportation, all rights reserved.)

Five

Destination Pacific Coast Highway

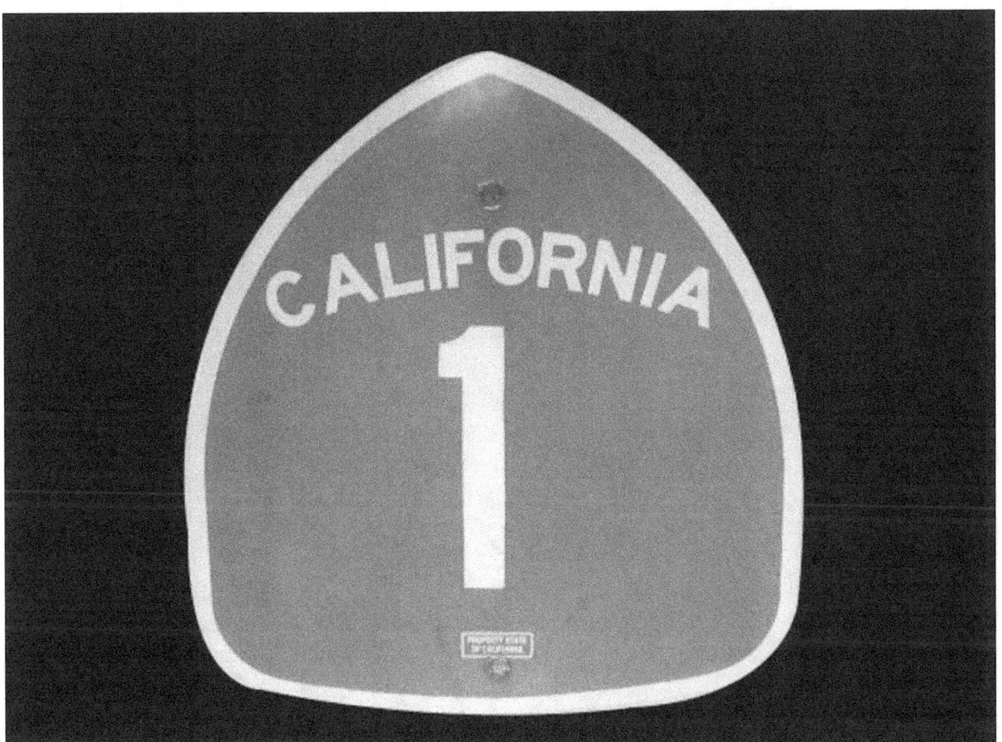

Although California's coast highway has several names, including Pacific Coast Highway, Cabrillo Highway, and Scenic Highway (to name a few), and although it is designated in certain areas as the 5 Freeway, US Route 101, and Highway 1, it is still most commonly known as the Pacific Coast Highway. The familiar green "California 1" sign with white lettering is associated with the coast highway.

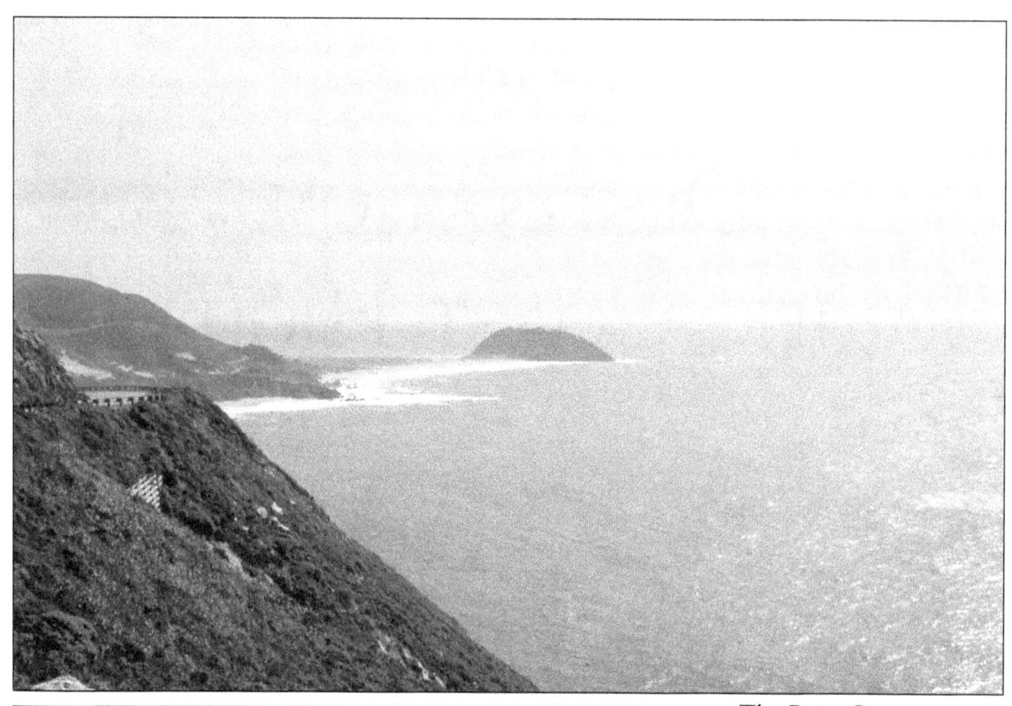

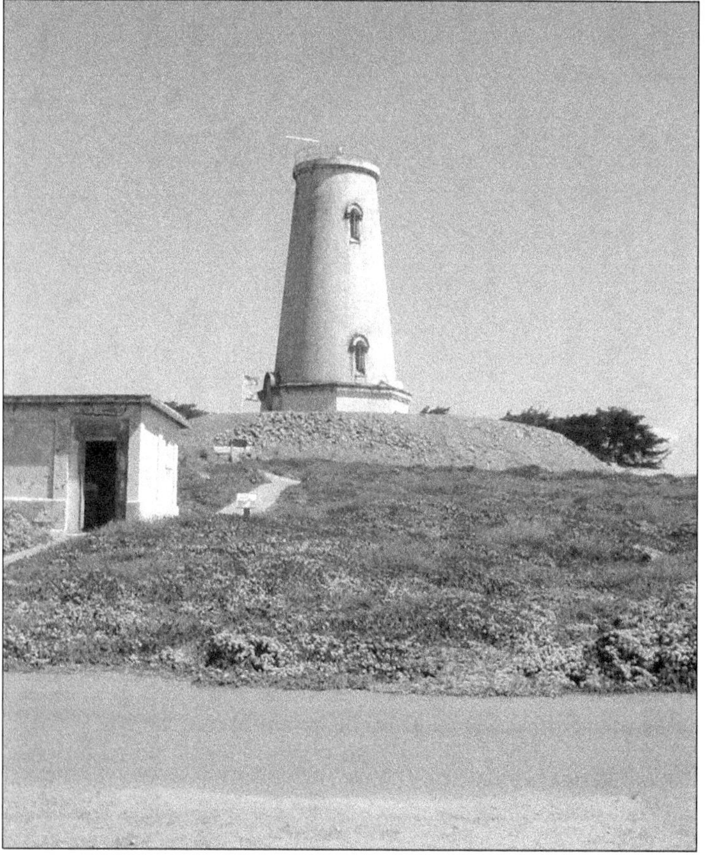

The Point Sur coast is pictured here as seen today. The picturesque, rugged landscape juts out into the ocean and has been the site of many shipwrecks, including the sinking of the USS *Ventura* in 1875. This prompted the need for a lighthouse on the point. (Robert C. Pavlik.)

The historic Piedras Blancas Lighthouse is pictured as seen today without its upper three floors. It was originally 100 feet, but earthquakes damaged its structure. Its fourth landing, watch room, and lantern had to be removed. (Robert C. Pavlik.)

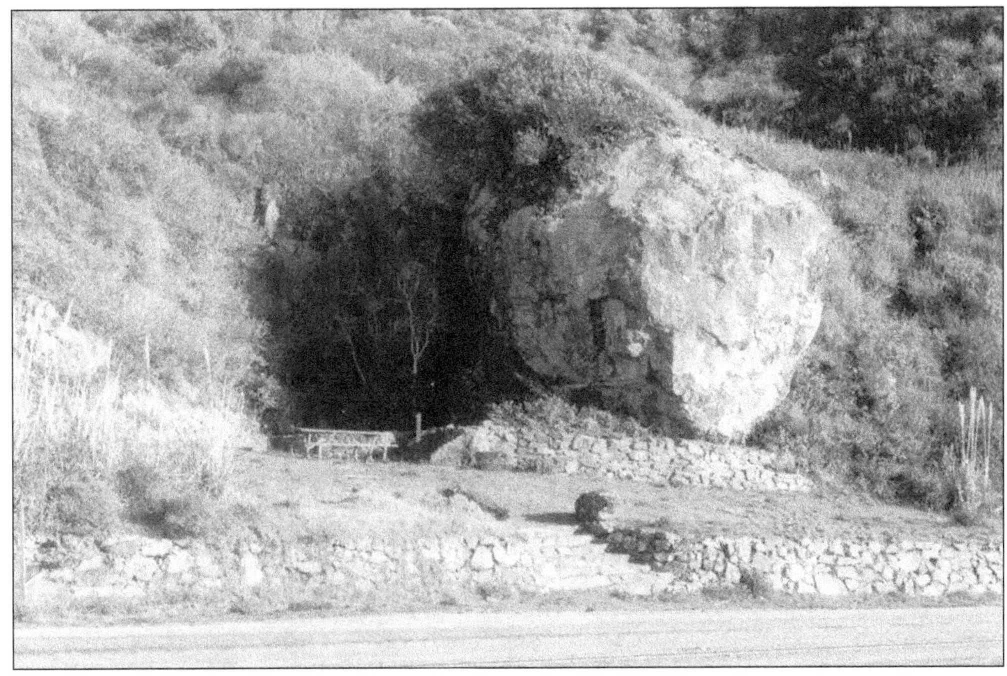

The Elmer S. Rigdon Drinking Fountain along the Big Sur Coast is seen here as it appears today. Rigdon was instrumental in acquiring state appropriation to build the coast road, but unfortunately passed away before the road was opened. The fountain was constructed south of Pfeiffer Big Sur State Park and dedicated to him on June 27, 1937, the same day of the celebration for the opening of the Big Sur Coast road. Below is another view of the Elmer S. Rigdon Drinking Fountain memorial. Its location nestled in the mountainside next to a huge boulder is symbolic of the construction of the coast highway. (Both, Robert C. Pavlik.)

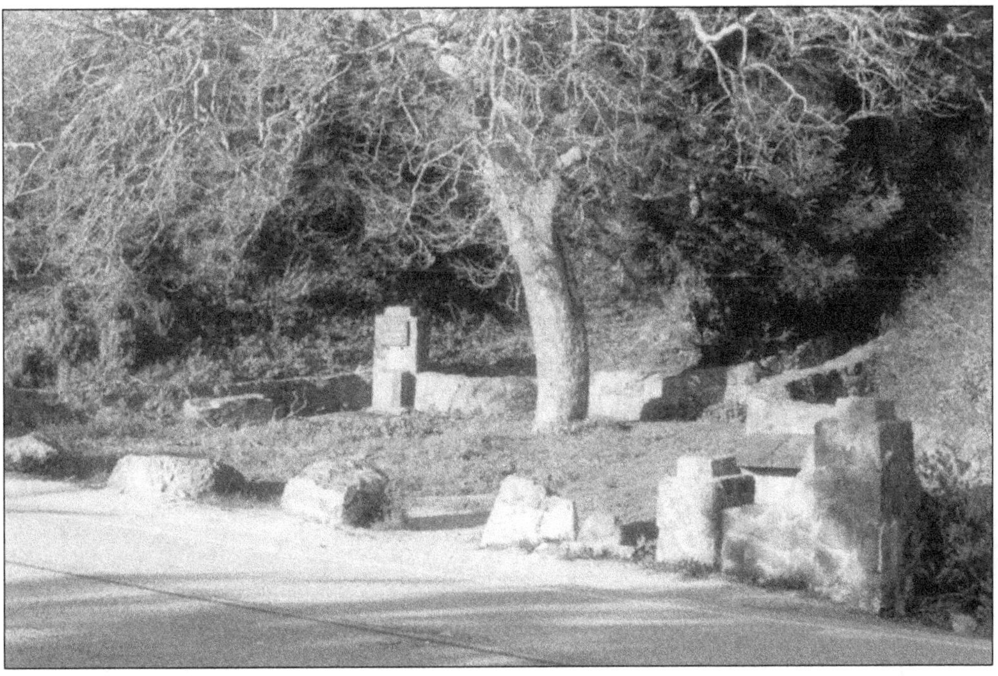

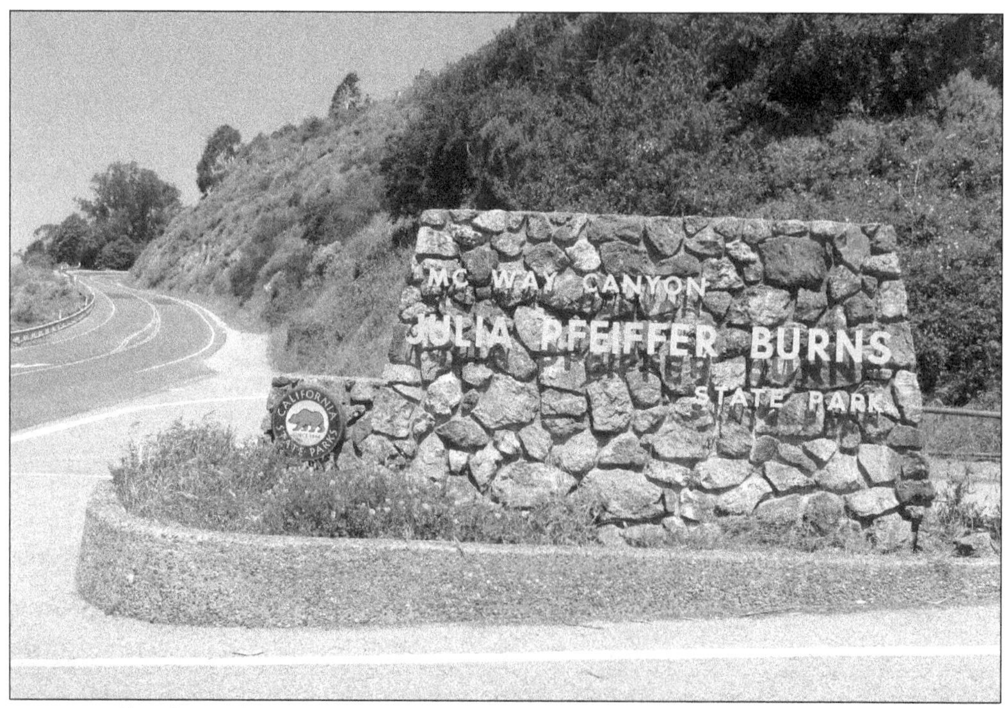

McWay Falls is located in the Big Sur region in Julia Pfeiffer Burns State Park. The picturesque 80-foot waterfall flows from McWay Creek directly into the Pacific Ocean. It is one of two waterfalls in California that cascade into the ocean. (Robert C. Pavlik.)

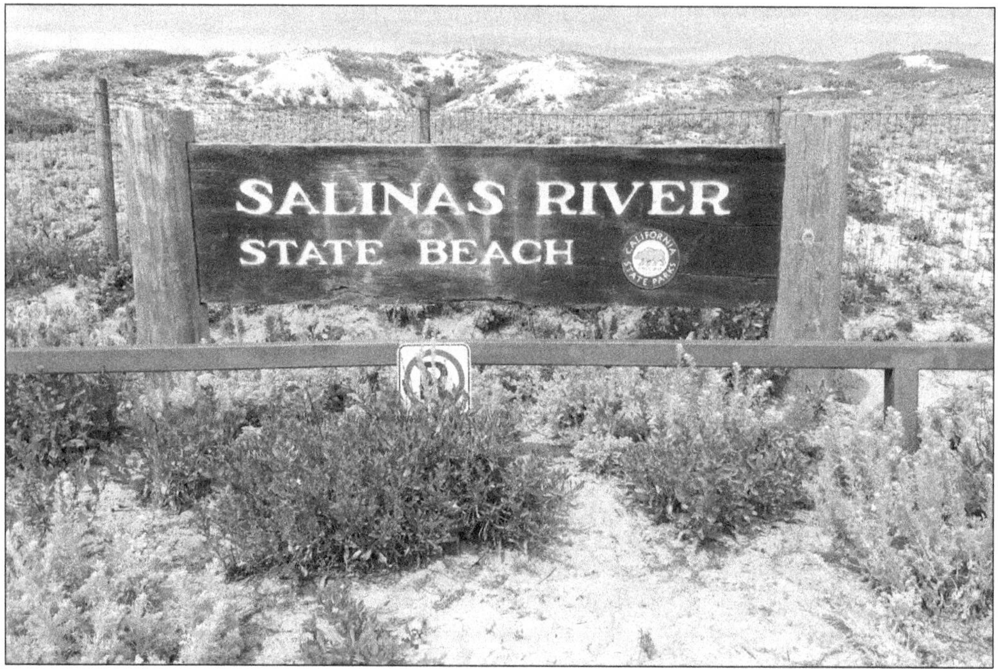

The Salinas River is the longest river on the central coast. It runs 175 miles and empties into the Pacific Ocean at Monterey Bay. Its beach is at the river mouth at Monterey Bay, and is long and wide with a backdrop of sand dunes. (Robert C. Pavlik.)

Pictured is the 80-foot McWay Falls at Julia Pfeiffer Burns State Park. The hand-split redwood Pelton wheel seen at the bottom right was built by Hans Ewoldsen in 1932. Ewoldsen was a foreman at the Saddle Rock Ranch, which supplied the first electric power in Big Sur. (Robert C. Pavlik.)

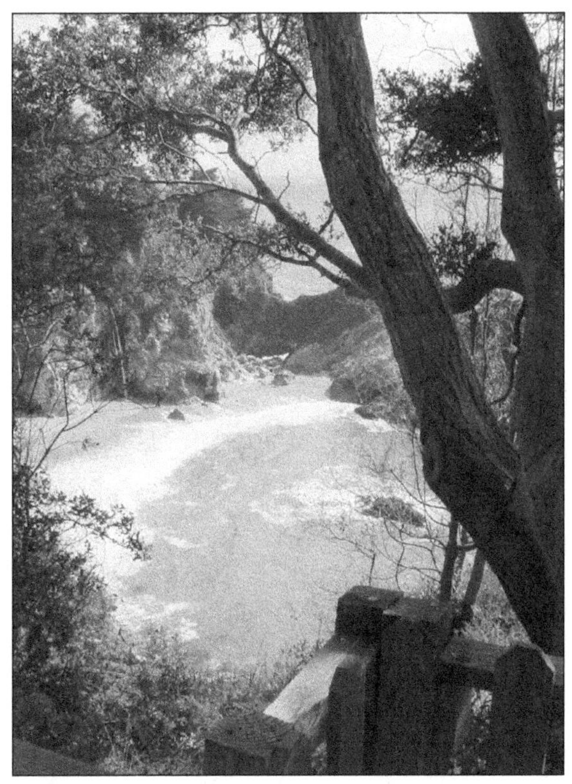

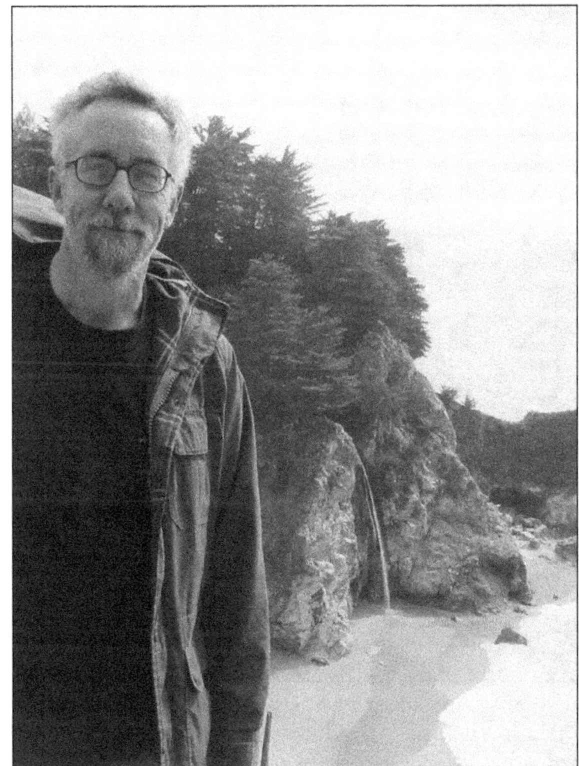

Robert C. Pavlik, a historian, author, and retired environmental planner for the California Department of Transportation, has journeyed to many, if not all, of the historic sites along California's Pacific Coast Highway and has generously provided many of his personal photographs for this book. (Rayena Pavlik.)

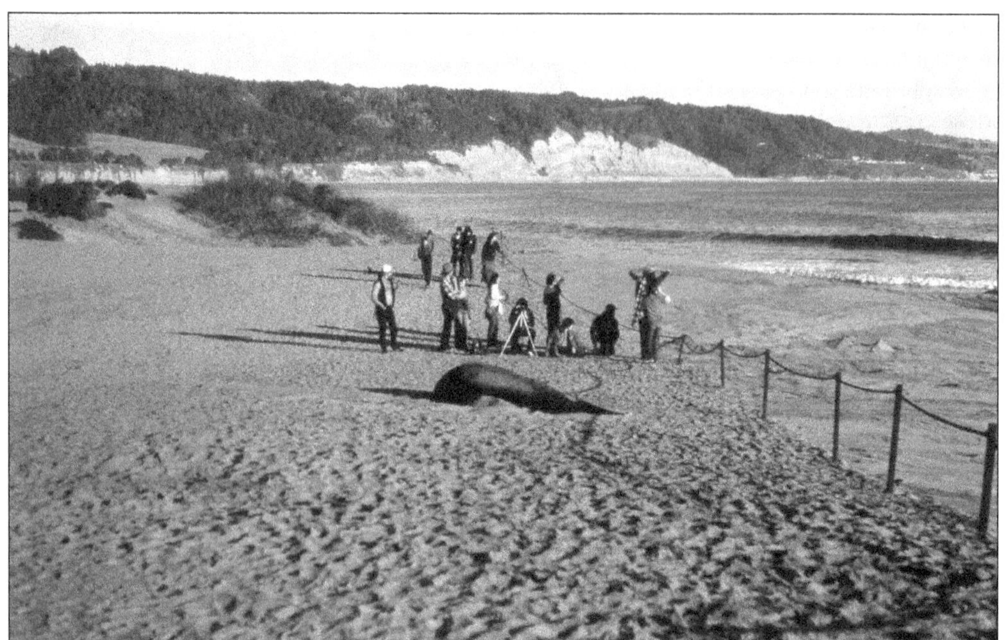

A visitor takes a picture of a California sea lion burrowing its head in the sand at Santa Cruz. During breeding season, sea lions gather on rocky shores. When not breeding, they gather at wharves, marinas, or buoys. Sea lions are a common sight along the California coast. (Robert C. Pavlik.)

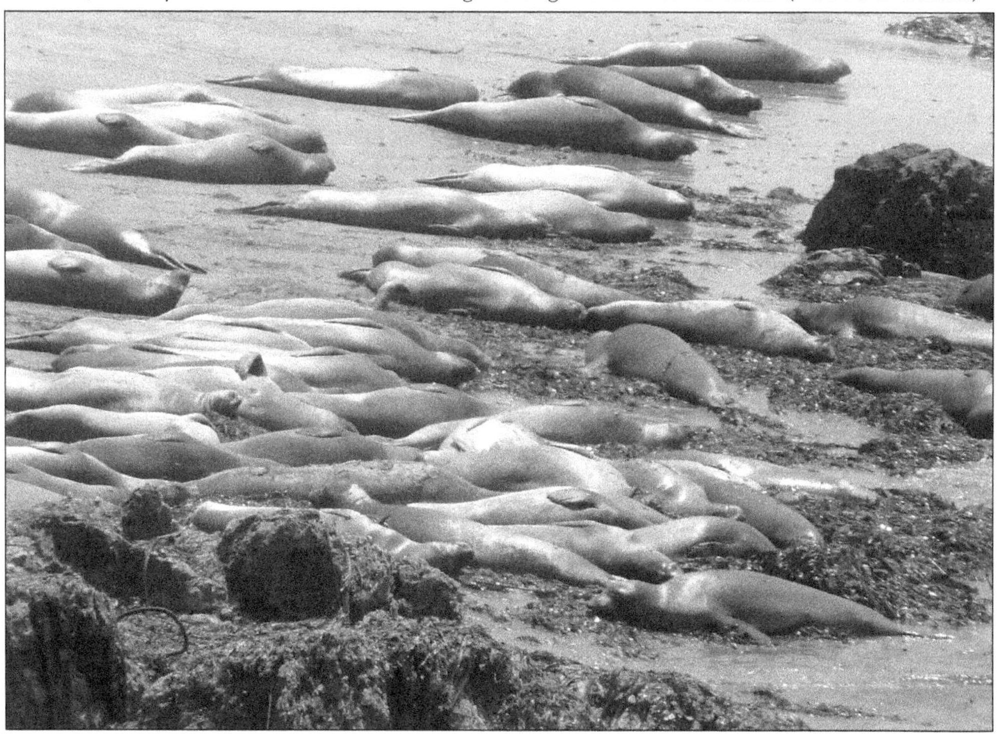

A large crowd of elephant seals bask in the California sun at San Simeon. Although elephant seals can be seen along the central and northern coast all year, they gather and bellow in large numbers on the beach during January, April, and October. (Robert C. Pavlik.)

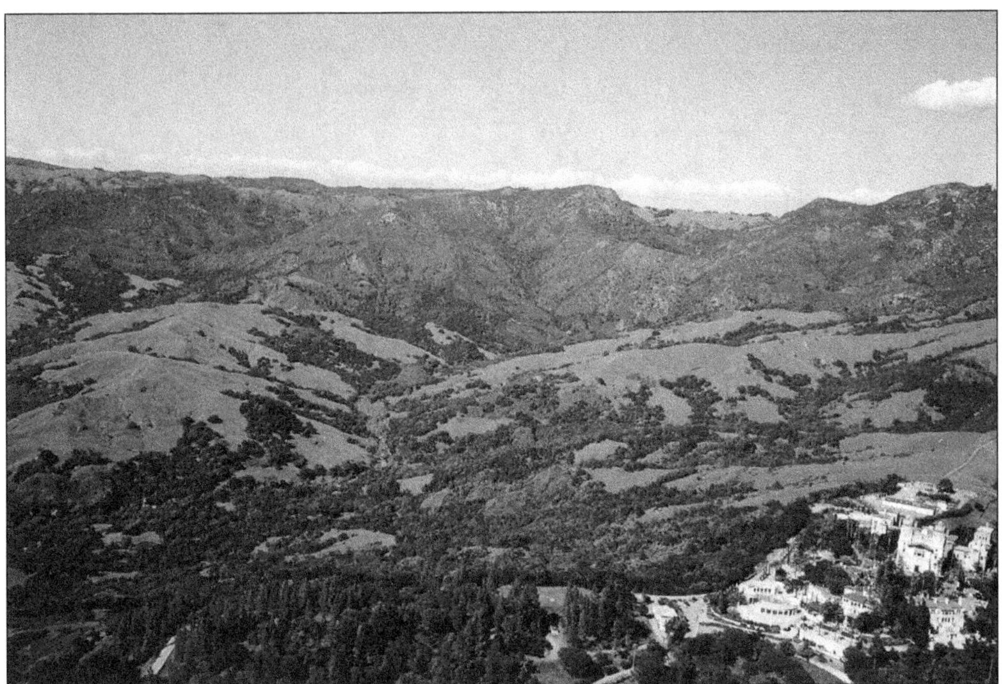

Above Highway 1 is Hearst Castle, former residence of newspaper magnate William Randolph Hearst, who formally named the estate La Cuesta Encantada—"the Enchanted Hill." Built on 250,000 acres, the castle is framed by the Santa Lucia Mountains. It is a National Historic Landmark and a California Historical Landmark. (Carol M. Highsmith collection.)

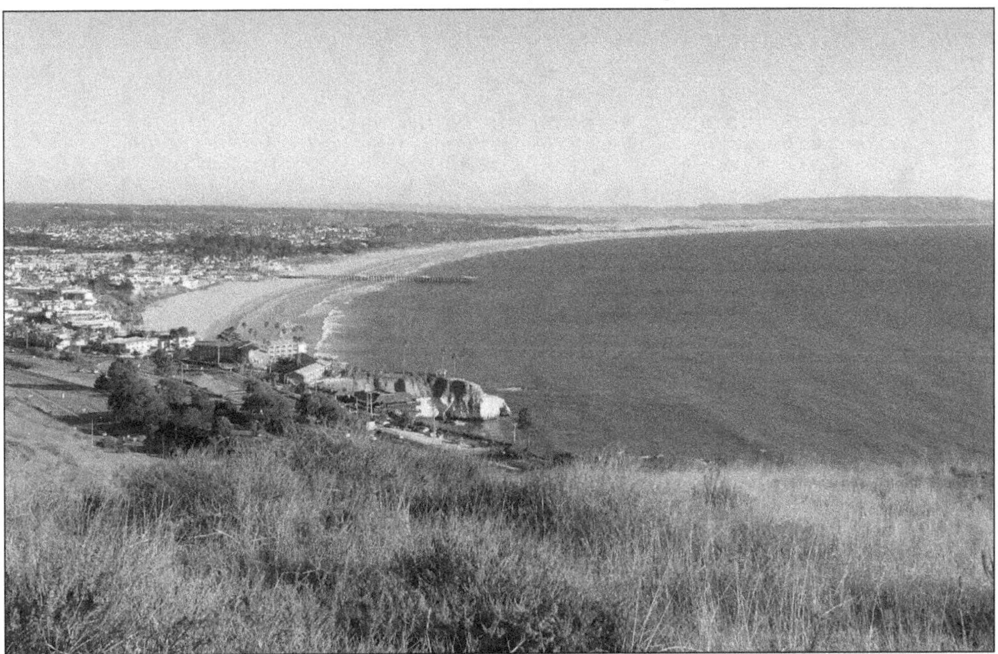

Off Highway 1 is the 17-mile Pismo State Beach. Part of the coastal Five Cities of Halcyon, Arroyo Grande, Pismo Beach, Grover Beach, and Oceano, it fronts three of the five cities. It is also known for its winter monarch butterfly grove. (Robert C. Pavlik.)

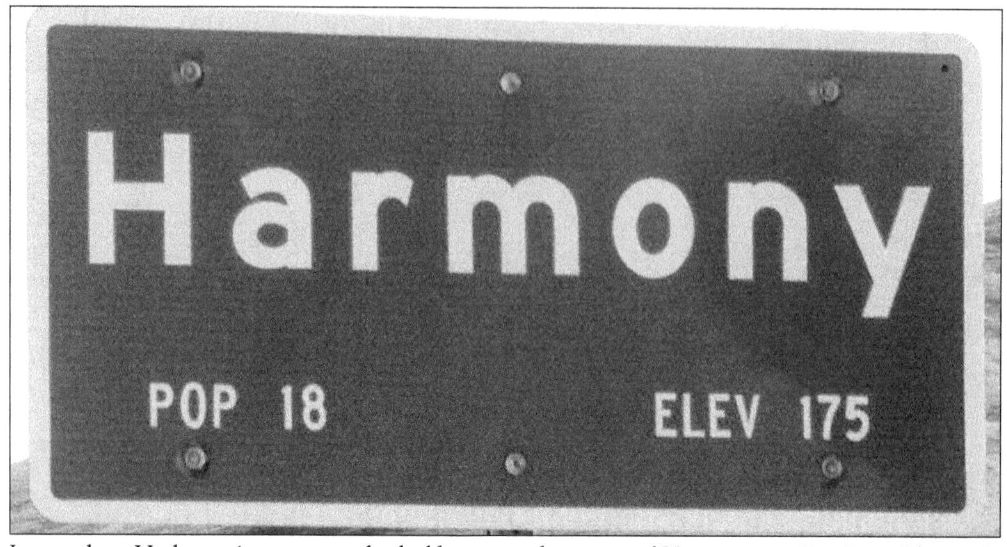

Located on Highway 1 on two and a half acres is the town of Harmony in San Luis Obispo. It was established in the late 1800s by dairy ranchers. There were rivalries among the ranchers, but eventually a truce was made, and the ranchers named the area Harmony to memorialize it. (Carina Monica Montoya.)

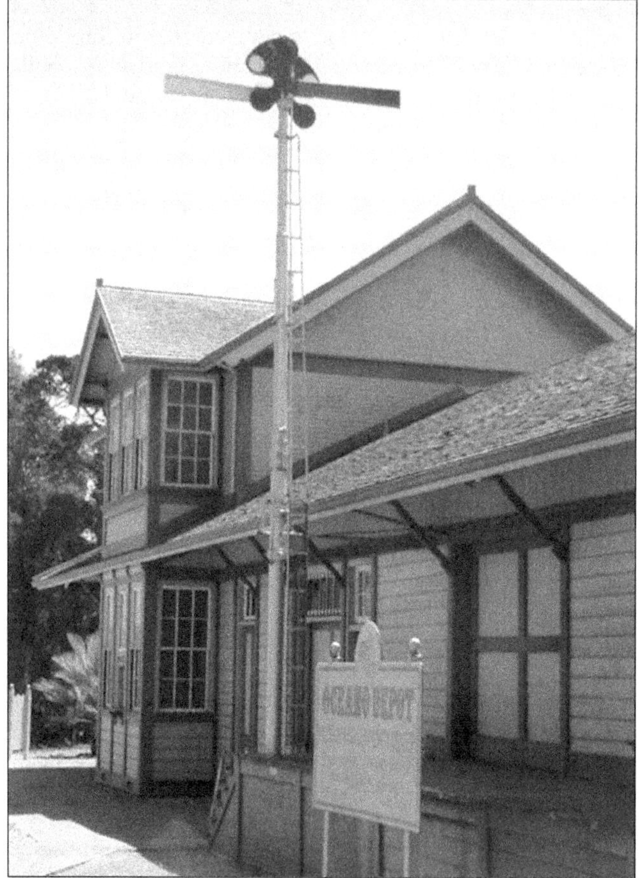

The Southern Pacific Railroad depot at Oceano is pictured as it is seen today. It replaced the original structure that was destroyed by fire in 1903. Established in 1896, it was once the central point for transportation and tourism. It was an important passenger and freight depot for San Luis Obispo. (Robert C. Pavlik.)

This plaque on the depot building memorializes its history. Nearby is Oceano's 1,500 acres of sand dunes. It is the only state park in California where vehicles may be driven on the beach. (Robert C. Pavlik.)

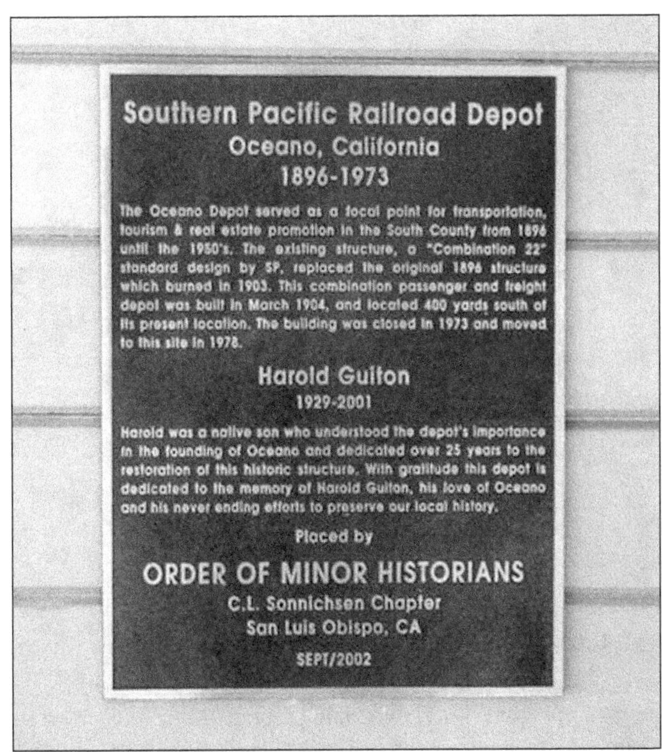

This is an aerial view of Vandenberg Air Force Base near Lompoc. Comprising 98,000 acres, of which 35 miles are rugged coastline, it is the only military base where unmanned government and commercial satellites are launched into orbit. It was formerly Camp Cooke (1941–1953), an Army training center for armored and infantry forces. (Vandenberg Air Force Base.)

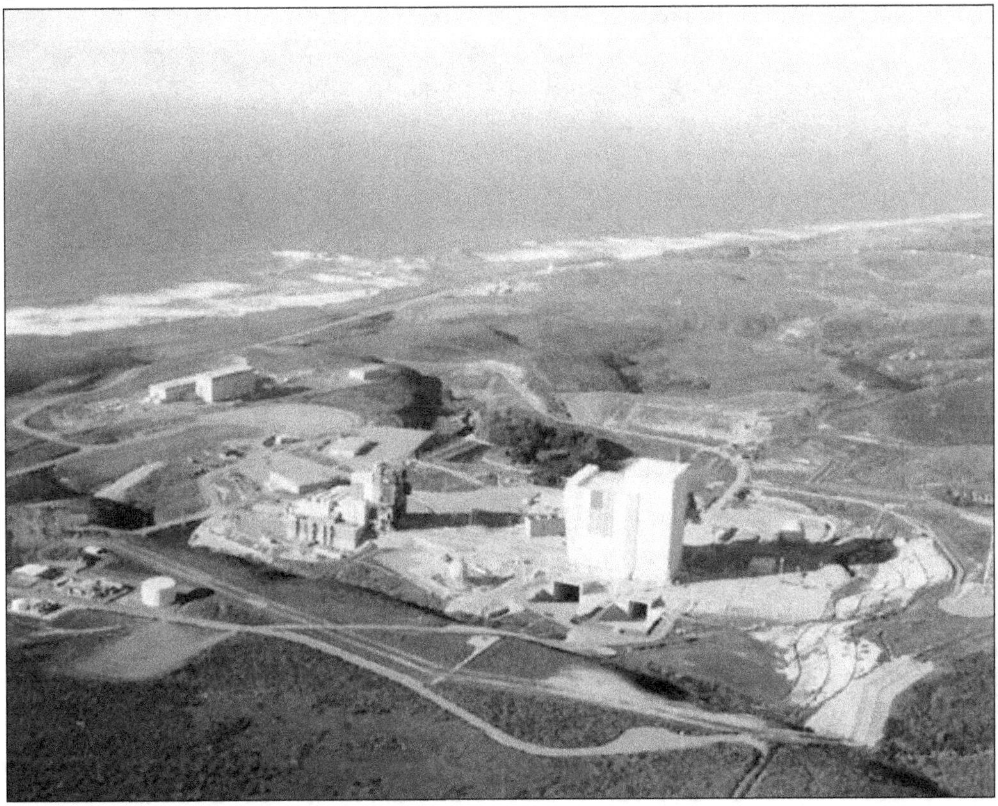

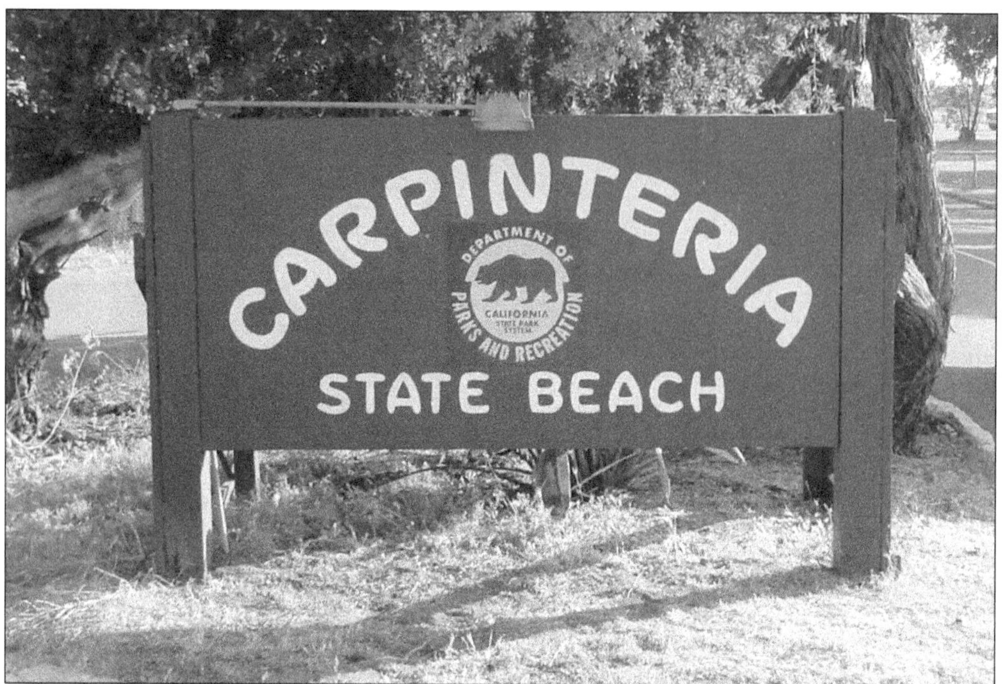

Carpinteria State Beach is a small, protected scenic beach off the coast highway in Santa Barbara County. Only a mile long, it is one of the most beautiful beaches in Southern California. Seals and sea lions gather in the winter months on the Carpenteria Bluffs. (Robert C. Pavlik.)

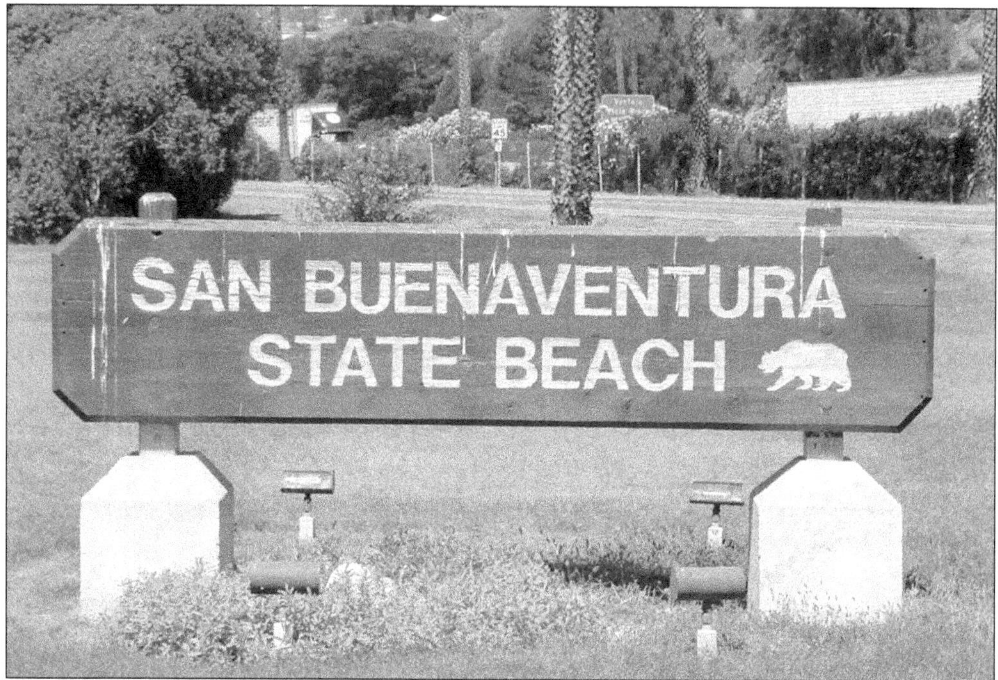

San Buenaventura State Beach runs along Highway 101. It marks the beginning of the famous Southern California beaches, where surfing and watersport competitions abound. Its two miles of waterfront stretch from Marina Park to the Ventura City Pier. (Robert C. Pavlik.)

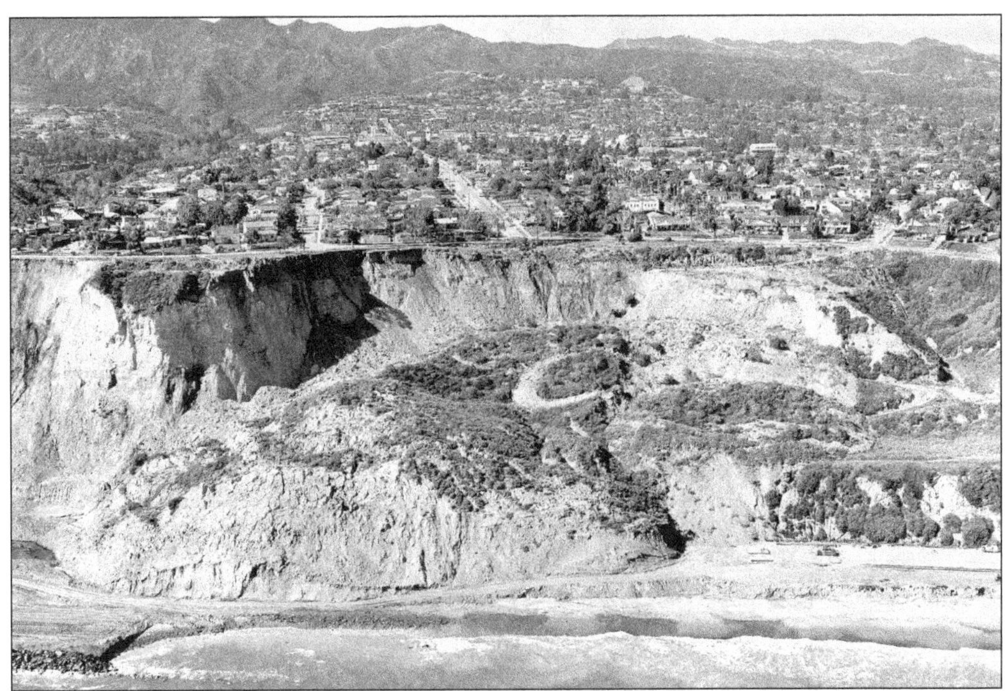

Pictured above is a 1958 landslide at Pacific Palisades. Massive amounts of earth hollowed a section of the mountain supporting the highway above. Excavations in building the road combined with natural erosion, corrosion, and salt water resulted in a high probability that a landslide would occur at some time. Below is another landslide at Pacific Palisades in 1983. The sides of the mountain had been cut away for housing development, which weakened the hillside and threatened structures alongside and below it. Fallen rock and debris are on the highway, as well as vehicles that are in danger of driving by an unstable area. (Both, copyright California Department of Transportation. All rights reserved.)

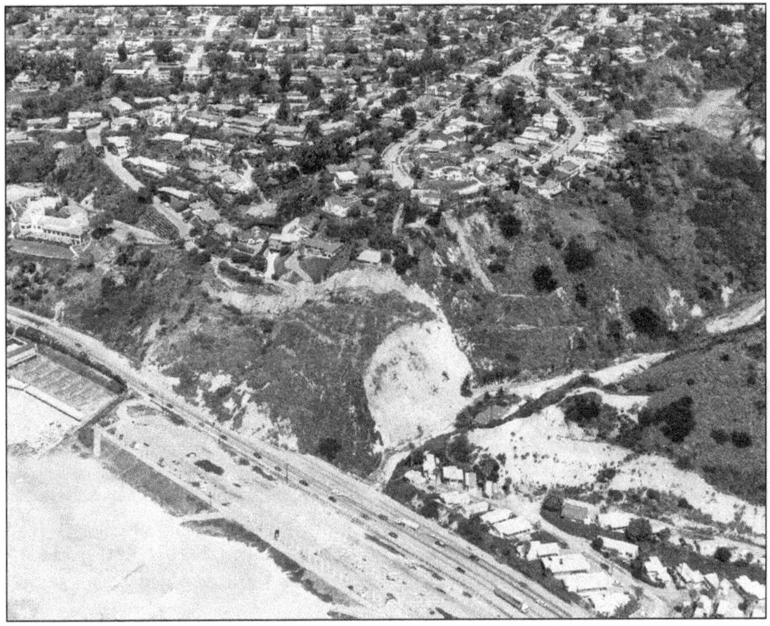

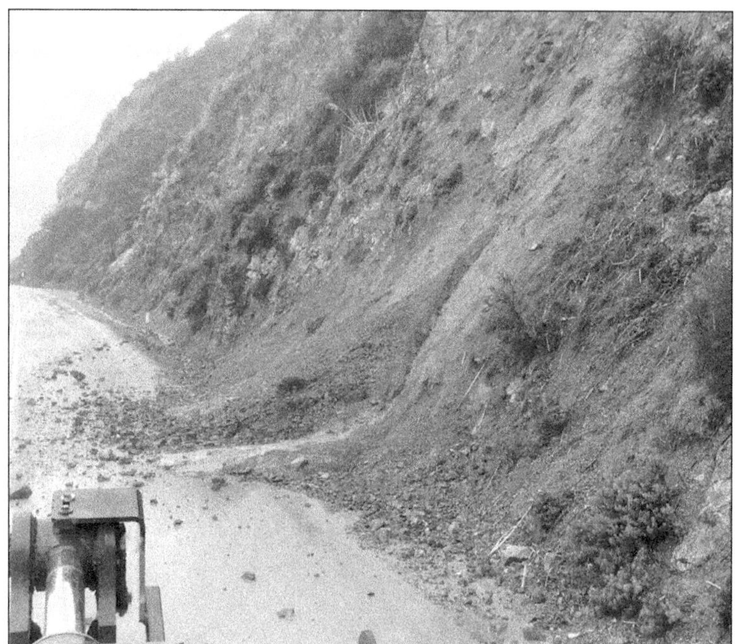

A small landslide at Anderson Canyon is seen here in March 2017. Record-breaking torrential storms hit Northern California and dumped several inches of rain that saturated mountains and hillsides, damaging a section of underground storm drain pipe. Although the slide did not damage the highway, eroded bluffs loom above the western edge of the road.

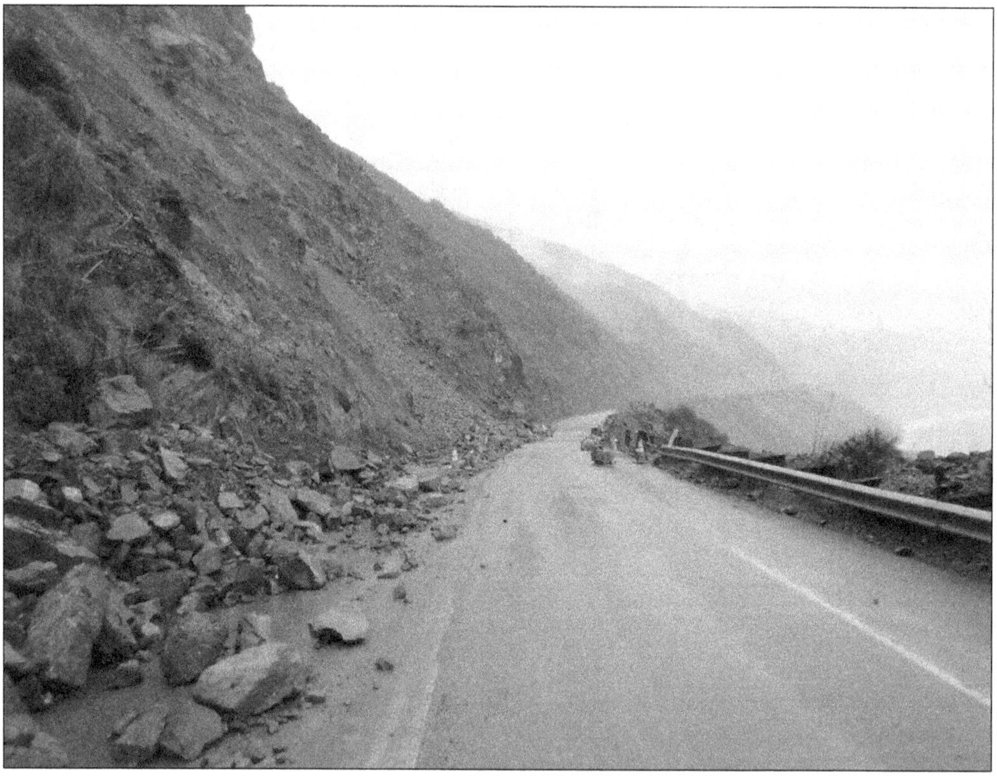

Pictured is the April 2017 Chimney Chute slide. Heavy storms triggered small, moderate, and devastating landslides along the Big Sur Coast. Fallen rocks can be seen scattered along the road, particularly at the base of the mountain along the road's shoulder. Larger rocks can be seen on the road next to the traffic cones.

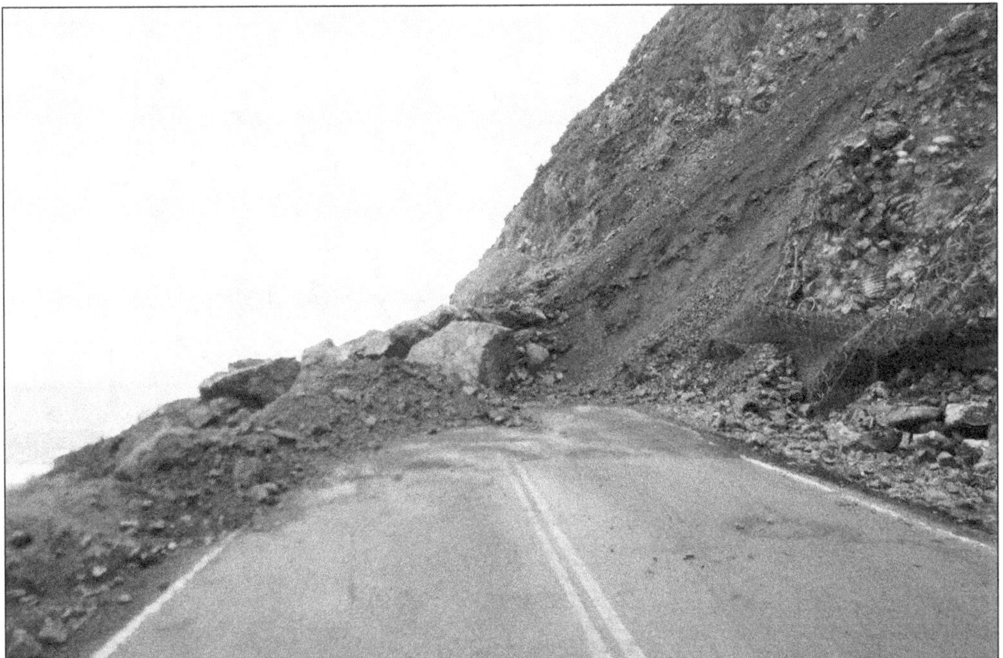

The Cow Cliffs slide in February 2017 is pictured here. Large rocks and boulders can be seen blocking this section of highway south of Lucia and Pfeiffer Canyon Bridge. Heavy rock slides such as this sometimes severely damage roads from the sheer force and speed of the falling rock hitting the pavement.

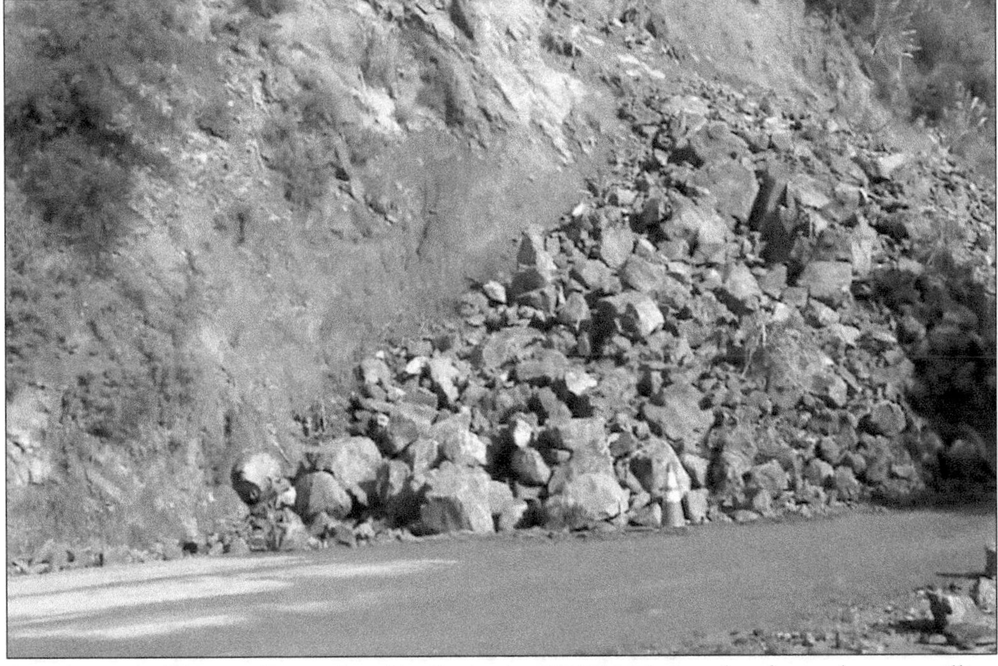

Seen here is a slide near Deetjen's in Big Sur in April 2017. Debris and rocks can be seen spilling onto the highway. Deetjen's Big Sur Inn is Big Sur's oldest hotel. Founded by Helmut Deetjen and his wife, Helen Haight, in the 1930s, it is listed in the National Register of Historic Places.

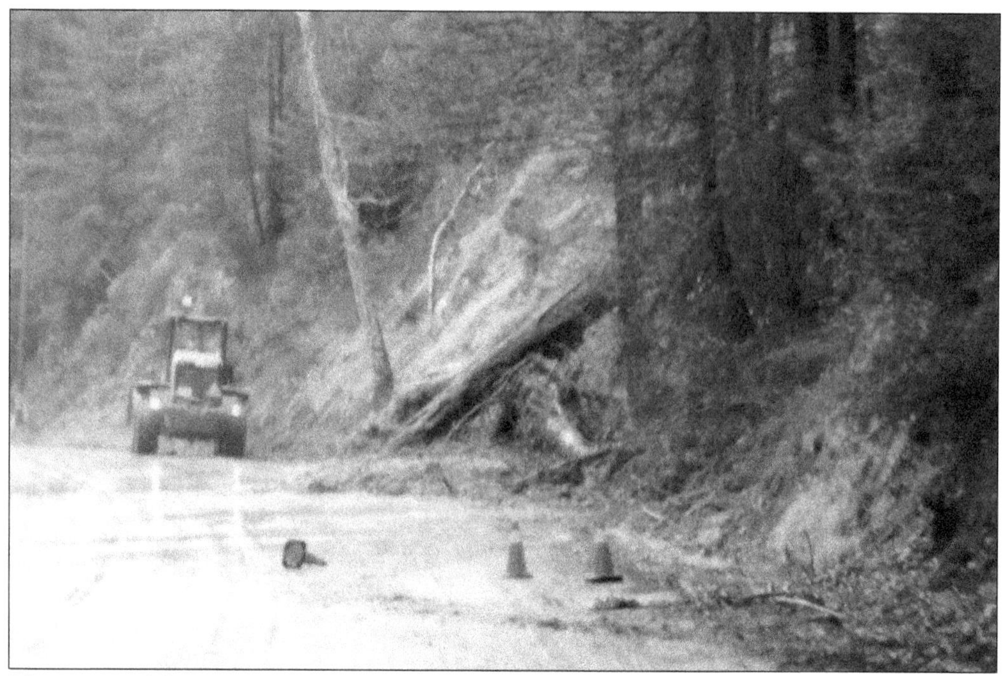

This image of the Fernwood Slide in January 2017 shows uprooted redwood trees that fell across the highway, causing road closures on the north end at Fernwood and Ragged Point. The landslide occurred in the Soberanes fire burn scar area, which weakened the soil, causing earth movement after heavy rains saturated the ground.

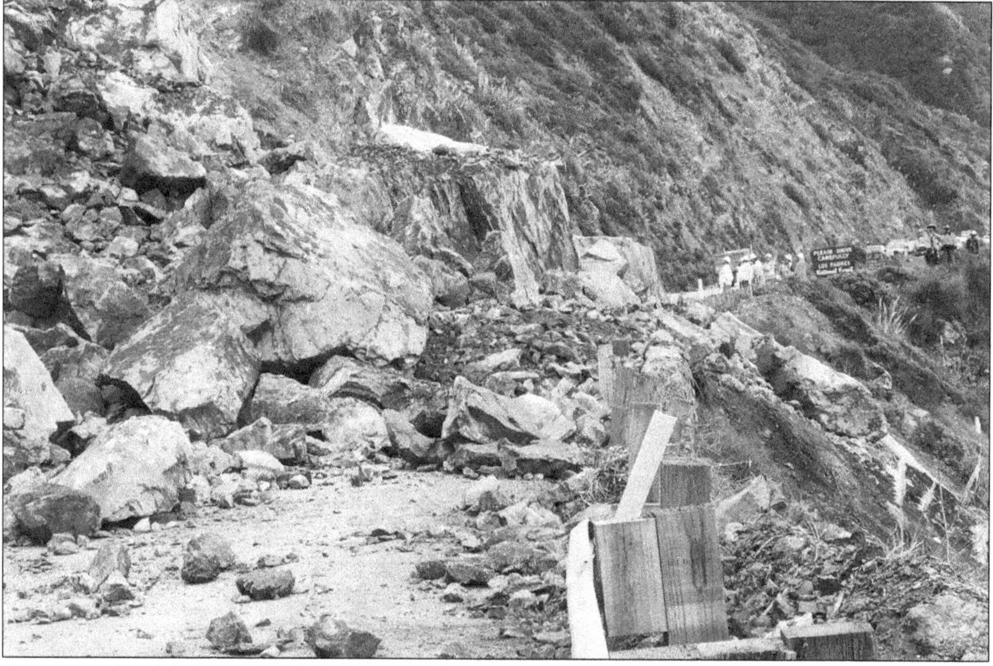

A massive landslide at Limekiln destroyed a section of Highway 1 in May 2017. The highway was buried under 40 feet of dirt and rock along a half-mile section of the Big Sur Coast. It was determined that the road closure along this section could last more than a year.

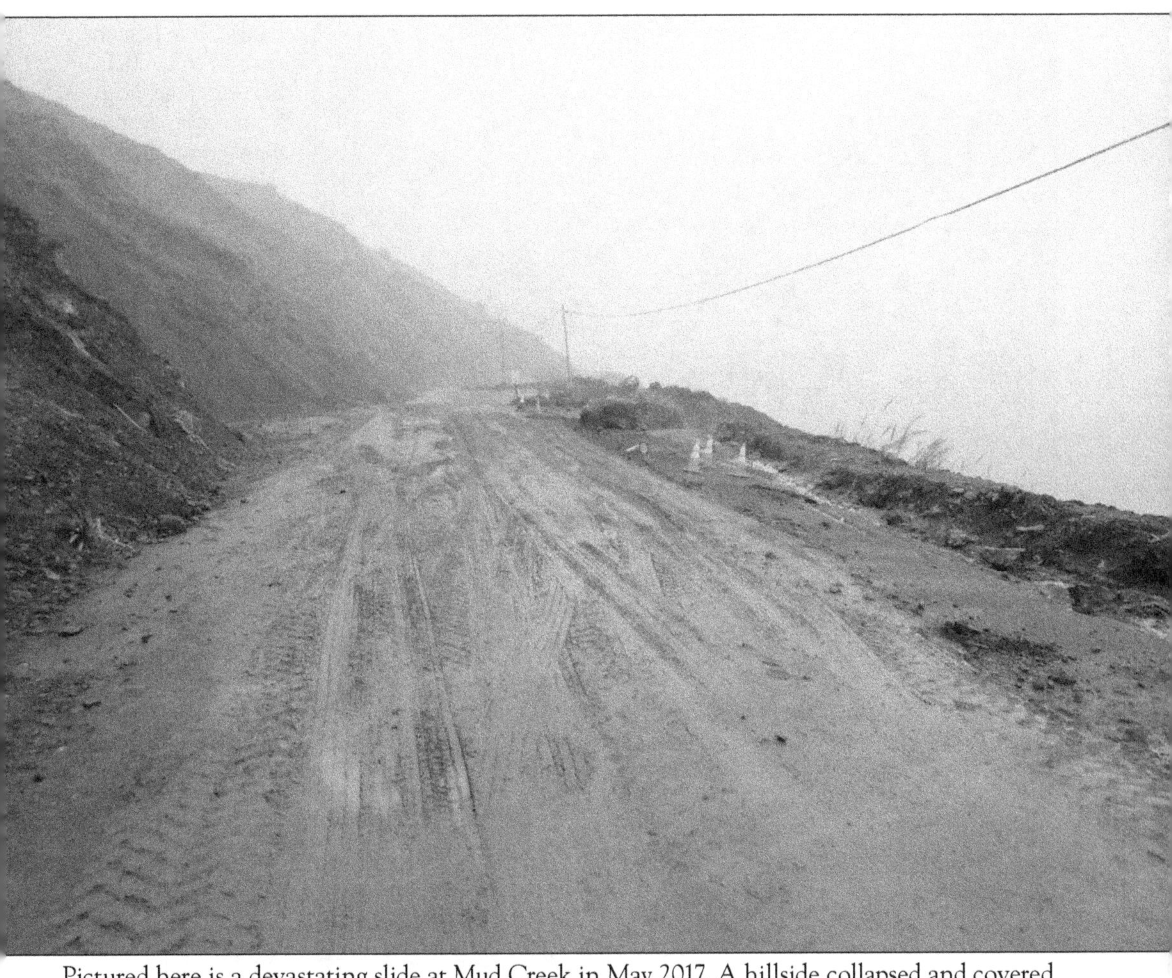

Pictured here is a devastating slide at Mud Creek in May 2017. A hillside collapsed and covered approximately a half-mile section of Highway 1. There were smaller slides as a result of heavy storms that caused one-lane closures, but this massive slide shut down the highway indefinitely.

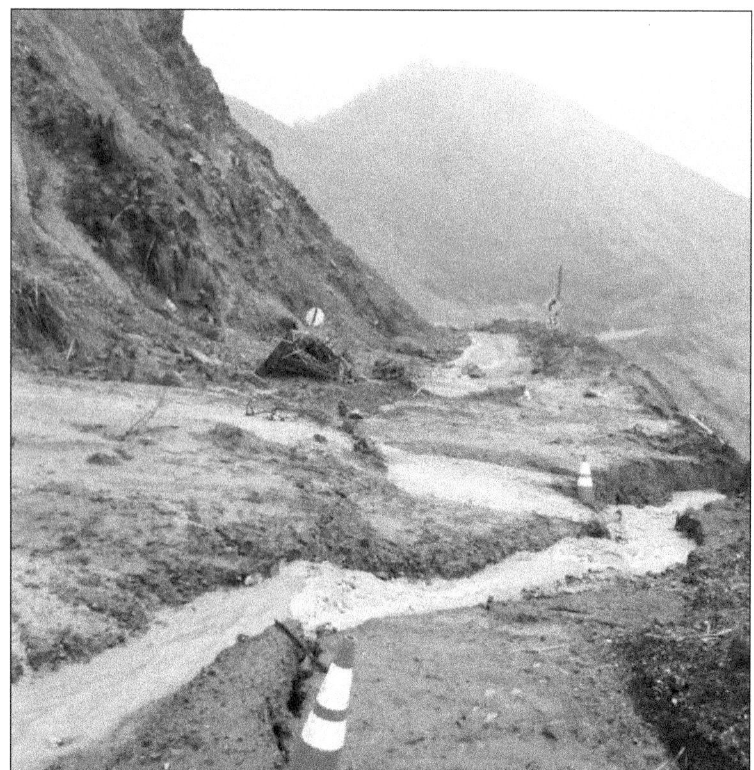

Paul's Slide is seen here in May 2017. Saturated ground from a wet winter plagued most of the central and northern coast. Work to clear dirt and rock was ceased because slide activity was increasing.

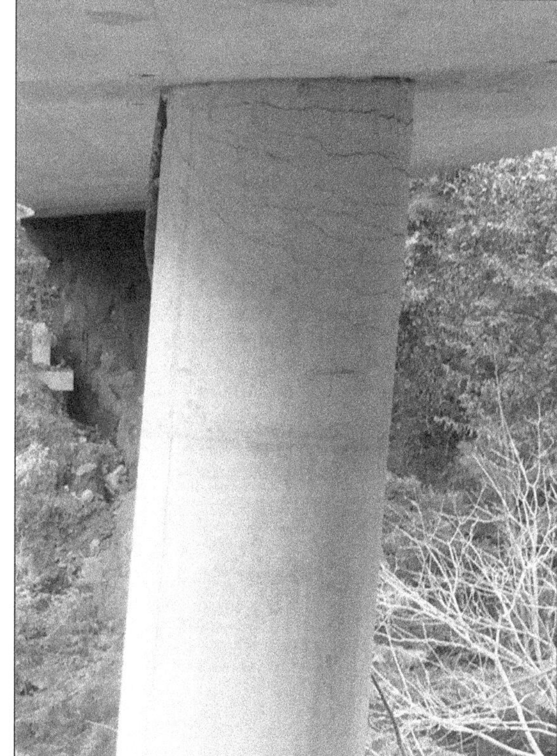

Pfiffer Bridge was damaged after strong storms and relentless heavy rains battered the central and northern coast in February 2017. The bridge began to sag as its support column began to crack and crumble. Residents and tourists in the area were stranded when the bridge failed, because the road north and south of the bridge was closed due to slides.

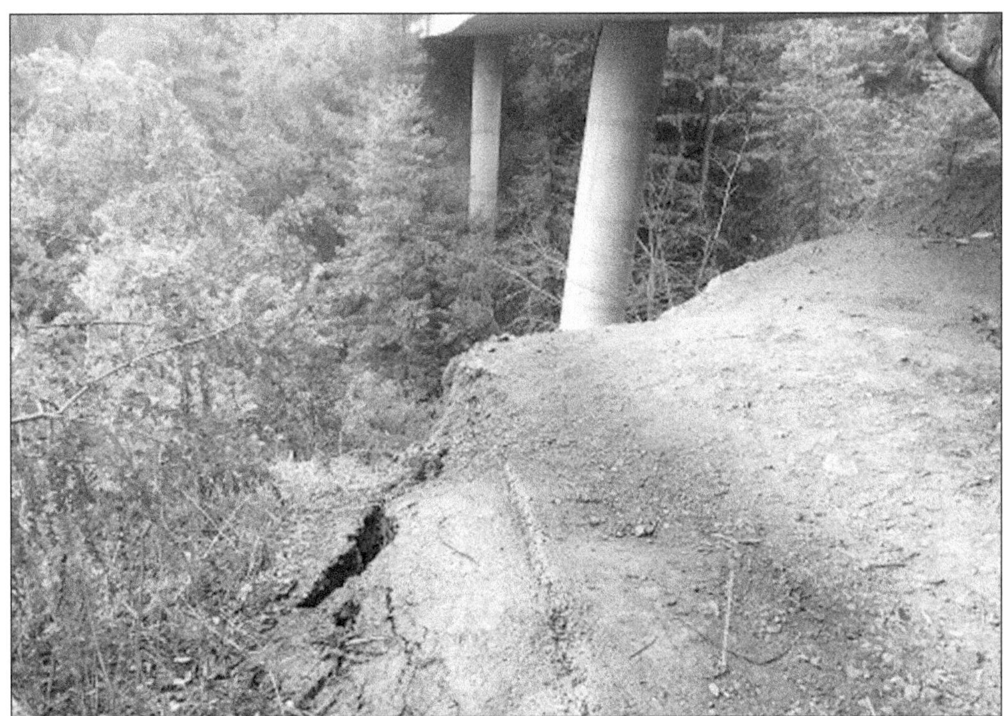

Seen here are the support columns of Pfeiffer Canyon Bridge that became damaged after the winter 2017 heavy rain and storms that hit the central and northern coast. One column is bent and significantly leaning to the side. The bridge had to be torn down, with much of the concrete tumbling down the canyon.

A large landslide of rocks and boulders in Los Padres National Forest is seen here. The section of road at the slide site is completely impassable, and fallen large rocks and boulders have destroyed the wood guardrails. Traffic can be seen backed up with nowhere to turn around.

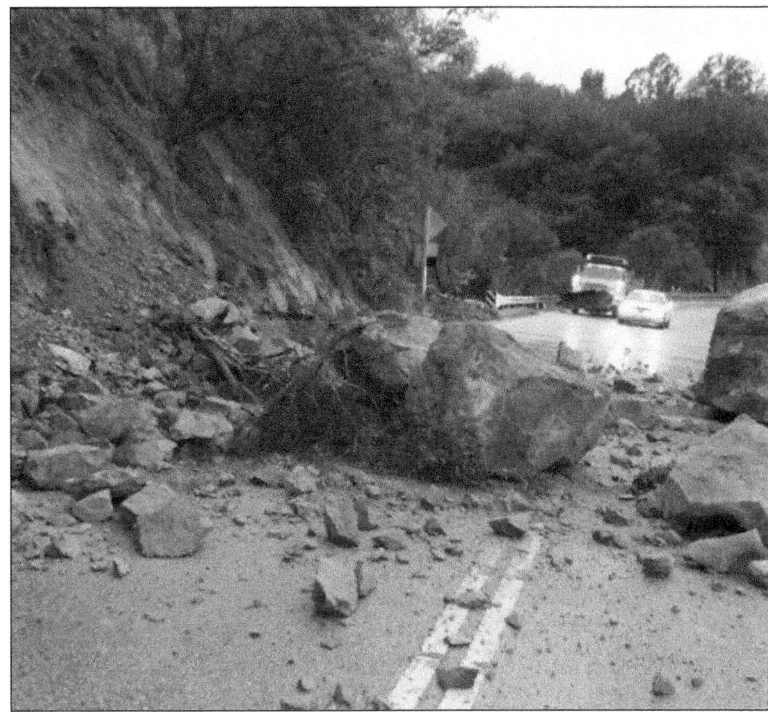

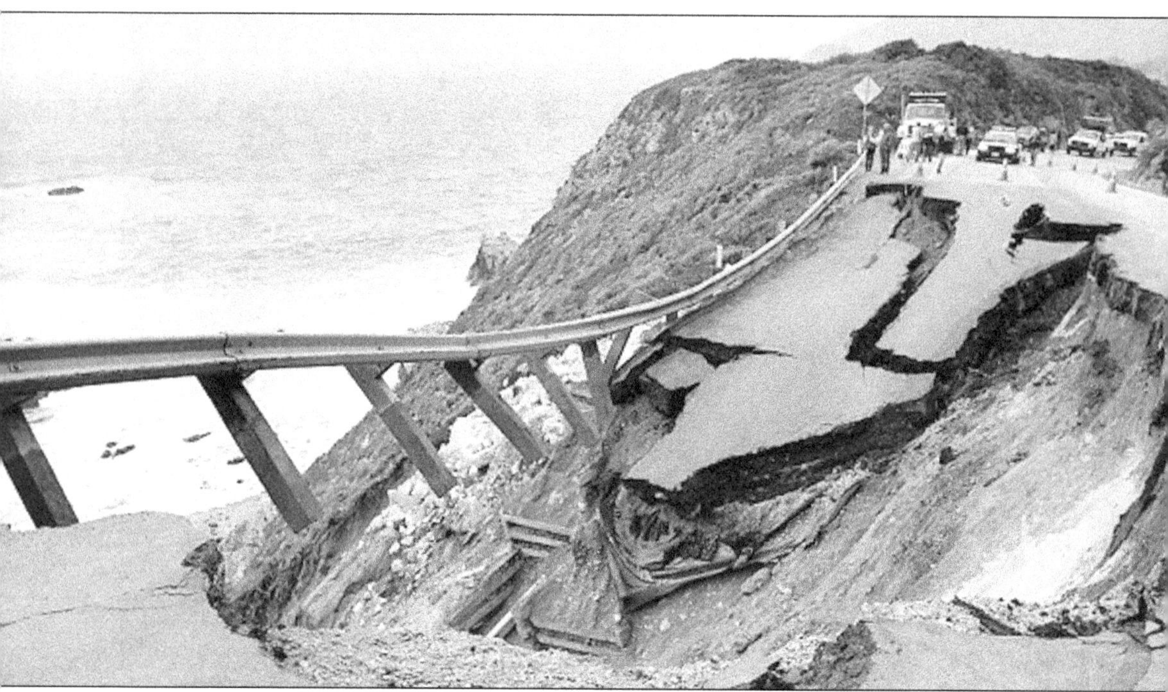

Another section of highway in Northern California along the coast also collapsed from heavy water saturation. The collapse of the earth under the road cracked the highway in large sections and bent the steel guardrails, sending large, cracked pieces of highway and debris into the ocean.

Heavy rains that saturate the ground can compromise the integrity of the road. Pictured here is a large crack that shows the ground giving away under the road. As the crack widens, it compromises other sections of the adjoining road. Major cracks can potentially become sink holes when water fills the hole. Below, more large cracks are visible on the road in Monterey. Cones are placed next to the areas that are separating. The many cones show how much of the road is giving out.

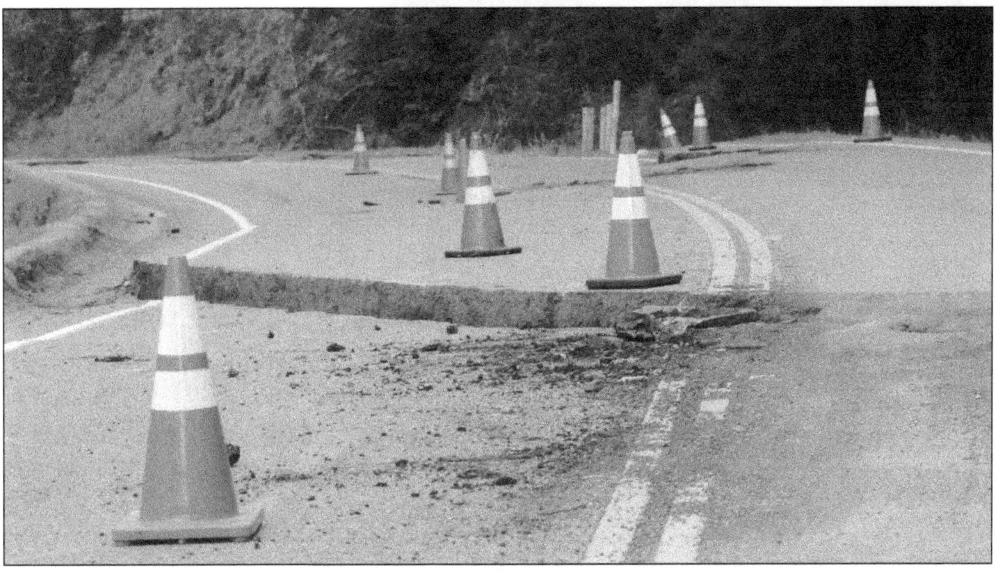

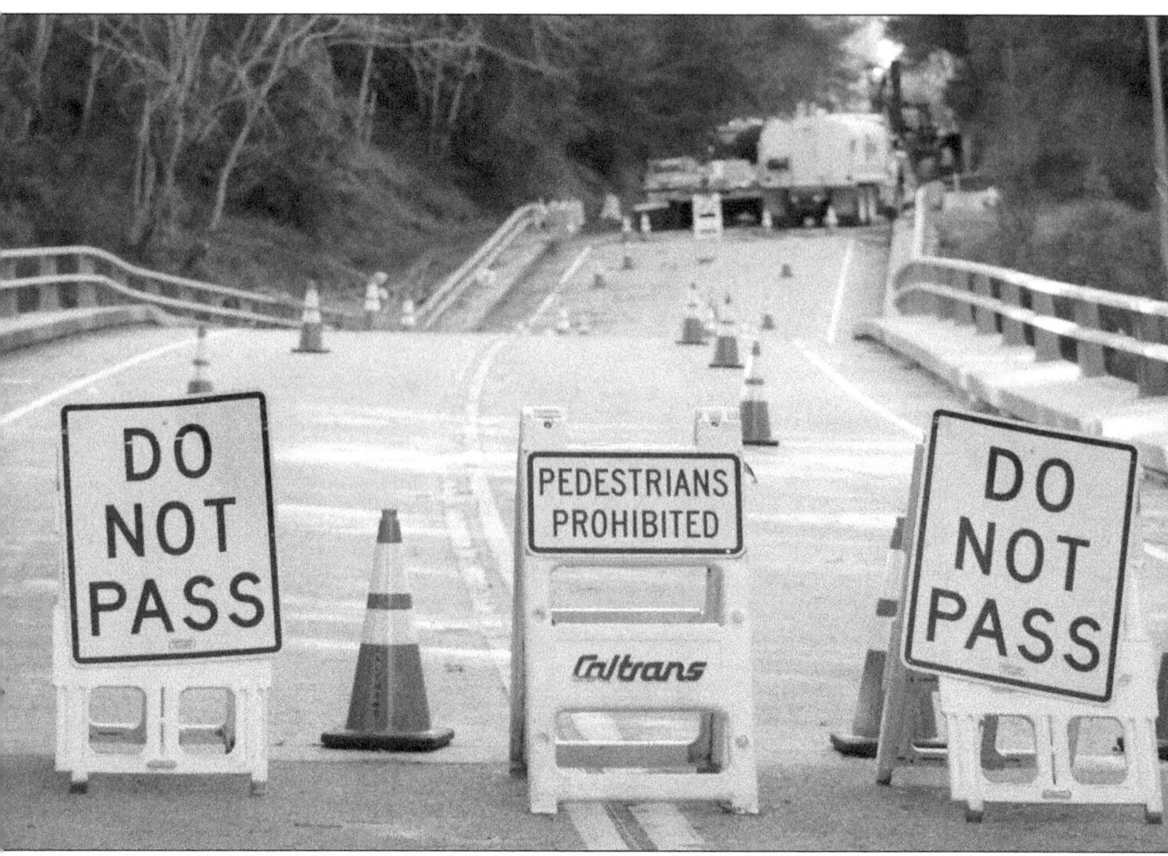

During the February 2017 storms, road closure signs were placed on just about every highway along the Monterey coast, particularly around the Big Sur Coast. These road signs were placed by the Pfiffer Bridge, which was collapsing, stranding motorists on both sides.

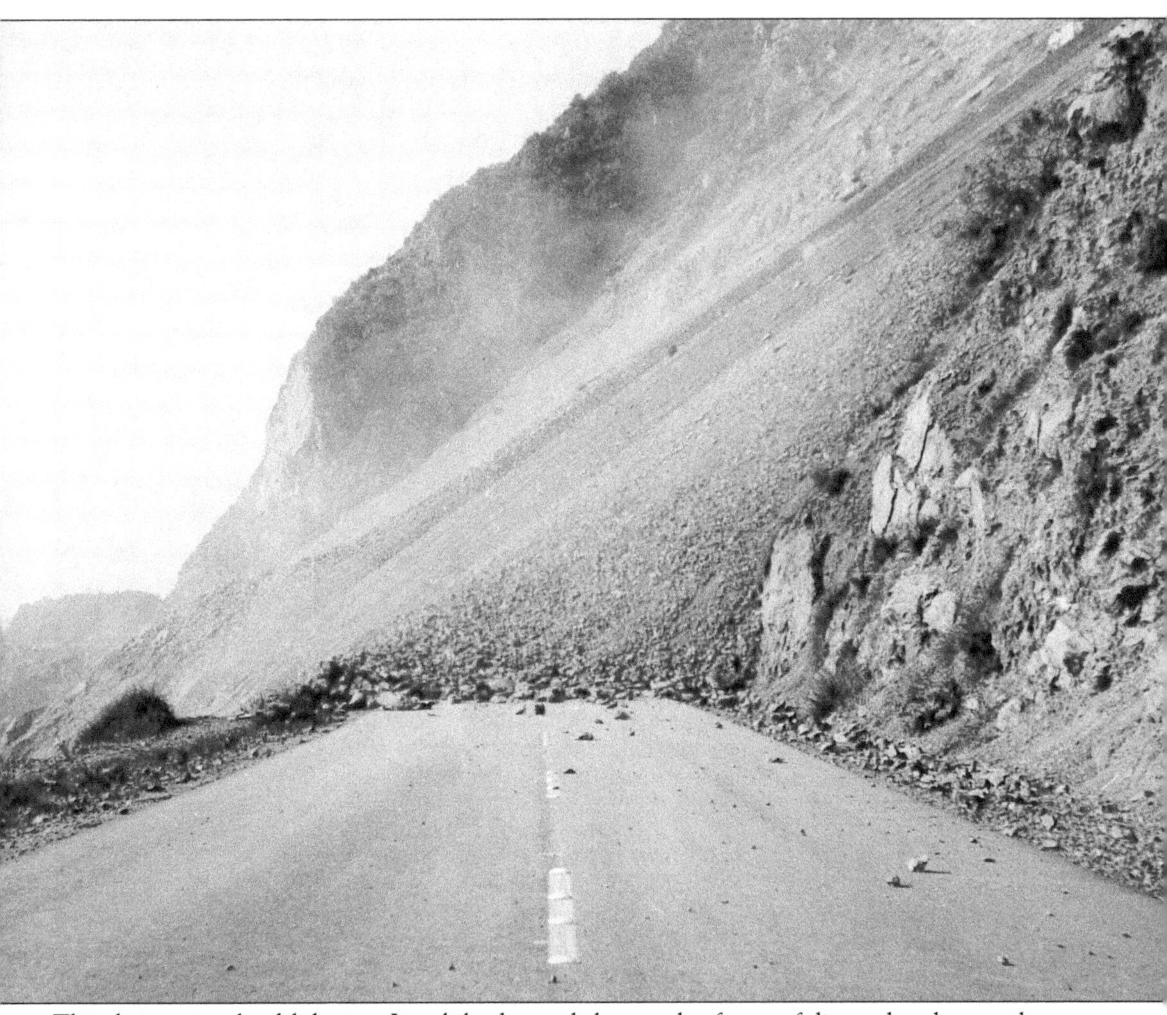

This devastating landslide near Limekiln dumped thousands of tons of dirt and rock onto the highway and over the cliffs into the ocean. The hillsides in this area are very unstable and prone to slides.

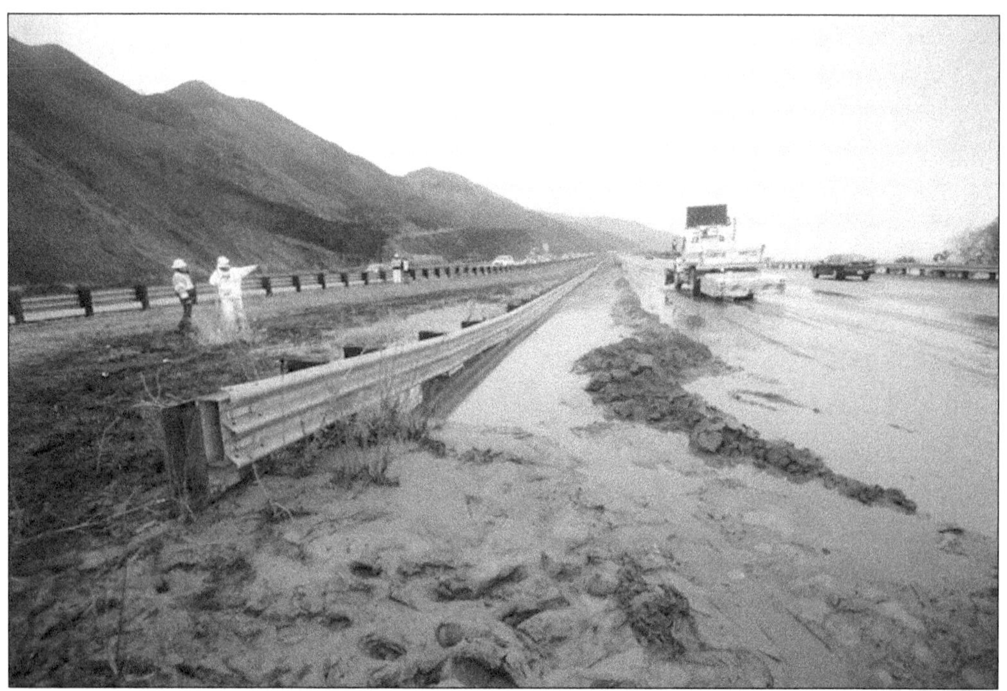

Mudslides can be just as devastating and damaging to the road. Mudslides are caused by water saturation in the mountains, causing the soil to loosen and give out. The force of a mudslide can uproot trees and move anything in its path.

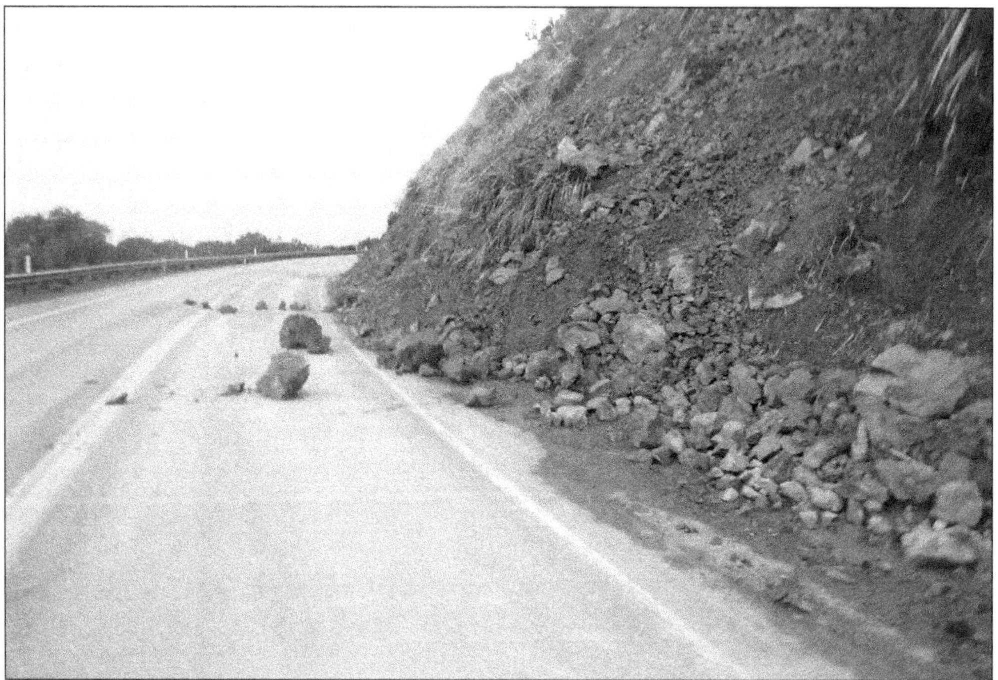

A small slide can still cause road damage and road closures. Pictured here is a slide along the Monterey coast. Dirt and rocks have spilled onto the highway and shoulder, making the section of road impassable. The danger is when the soil is unsettled and can continue to slide.

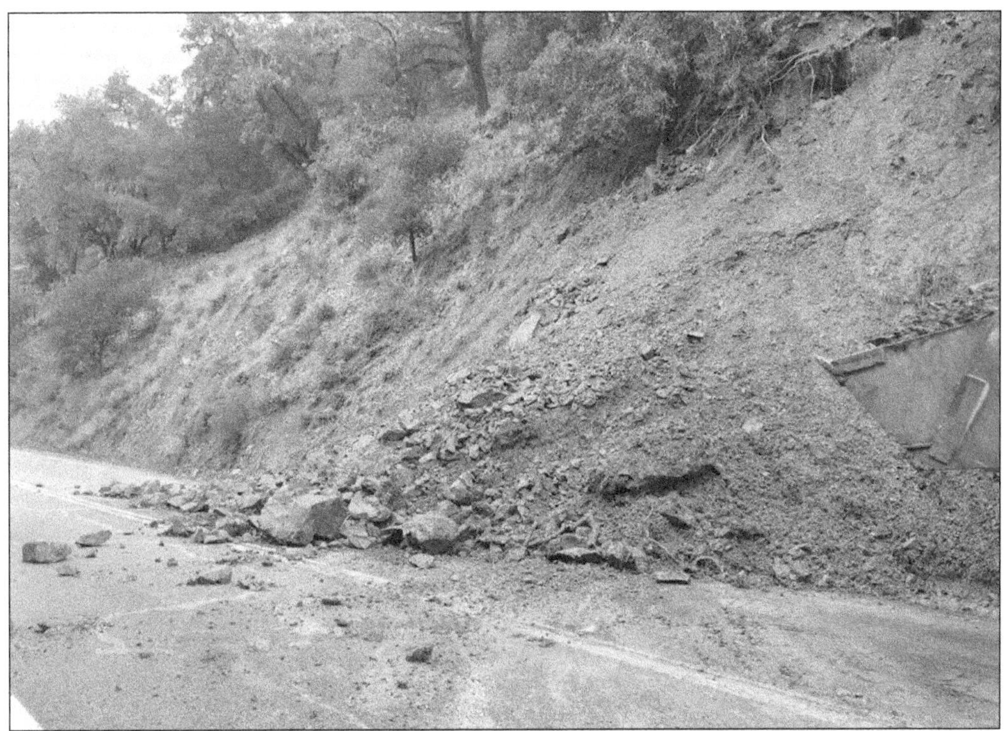

This small slide has spilled onto the highway. The hillside appears very loose and unstable. A heavy rain would likely cause this section to slide more rock and debris onto the road.

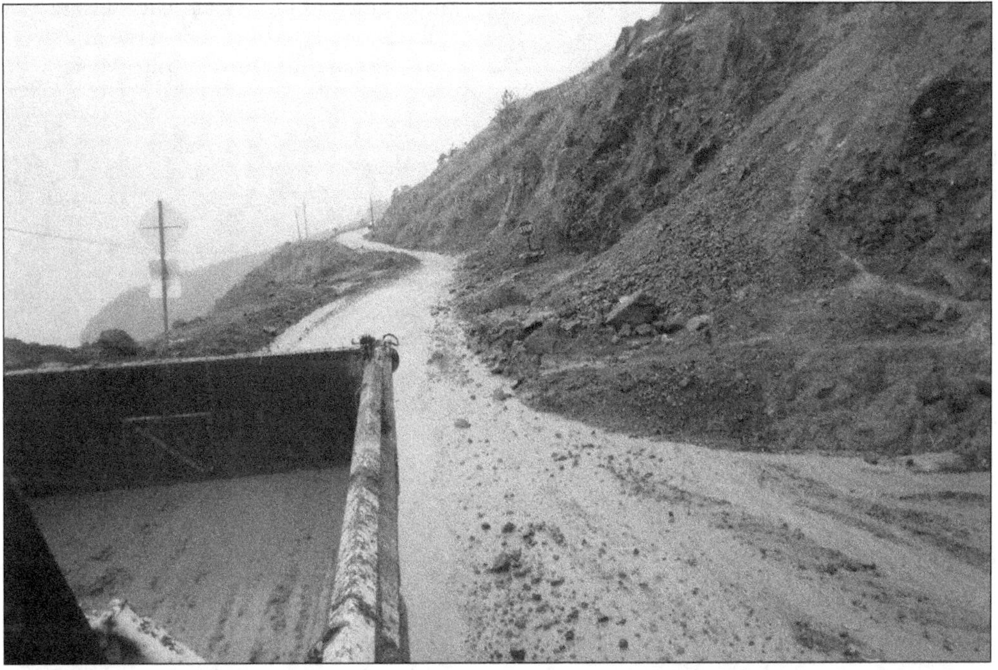

This slide has just about covered a section of road along the Big Sur Coast. The slide covers a large section of the road. This section of highway is narrow and sandwiched between the mountains and cliffs.

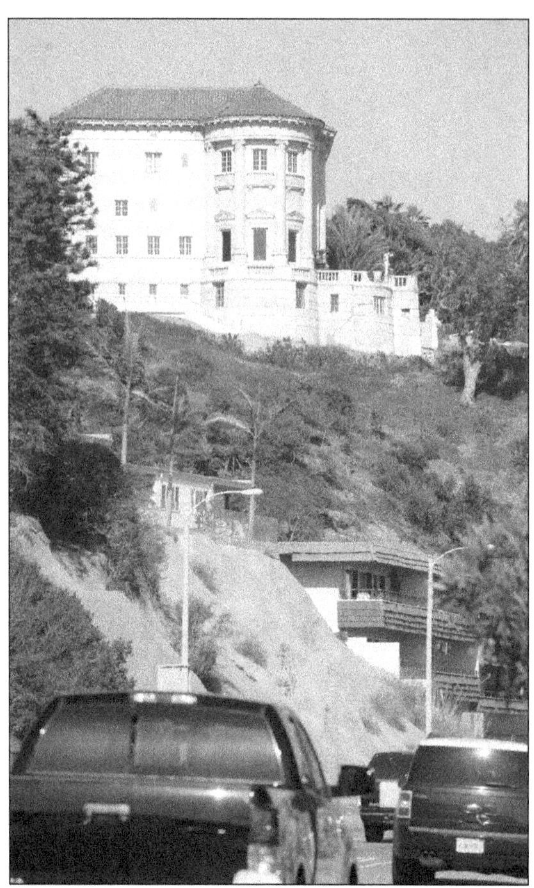

The Pacific Coast Highway is not all doom and gloom. The attractions along the coast bring people back time and again. The Getty Villa perched on a hill overlooking the Pacific Coast Highway straddles the border of Malibu and Pacific Palisades. (John Moss.)

The Pacific Coast Highway in Southern California is dedicated to Vietnam veterans. Other parts of the coast highway have different memorials, but they are all memorials to members of the military who served the country. (John Moss.)

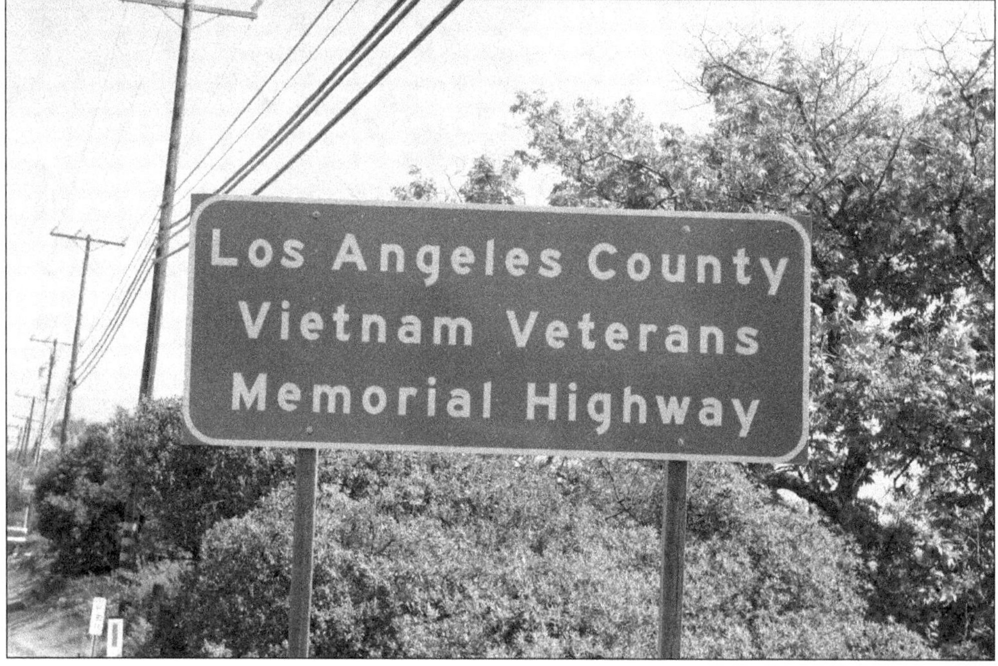

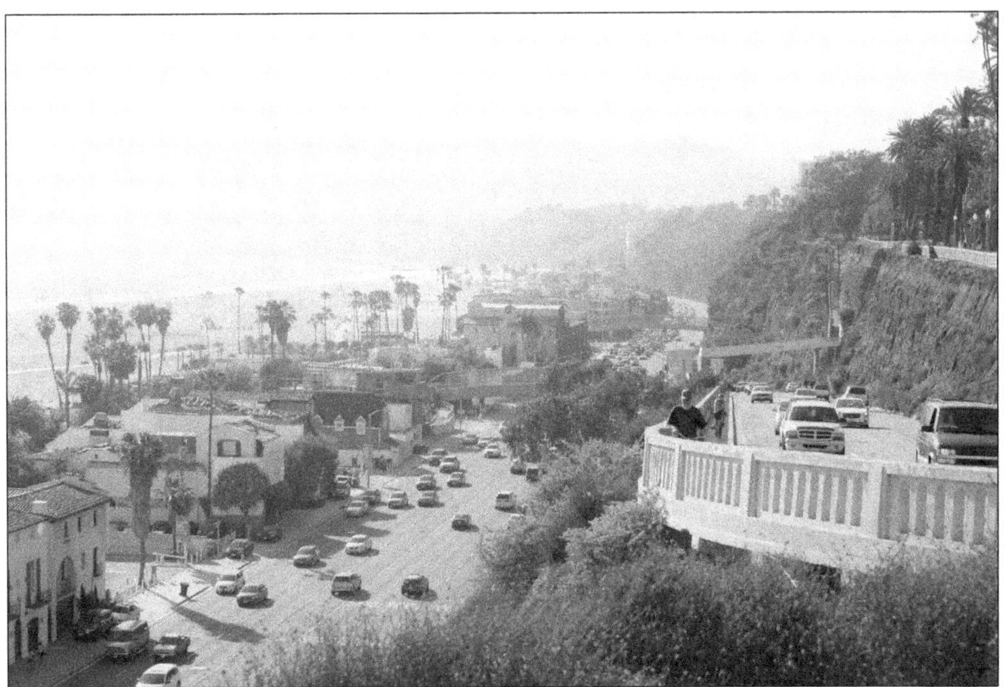

The California Incline in Santa Monica connects Ocean Avenue with Pacific Coast Highway. It was once a walkway in the 1800s for pedestrians to walk down to the beach. The incline is 1,400 feet and was built in 1930. (John Moss.)

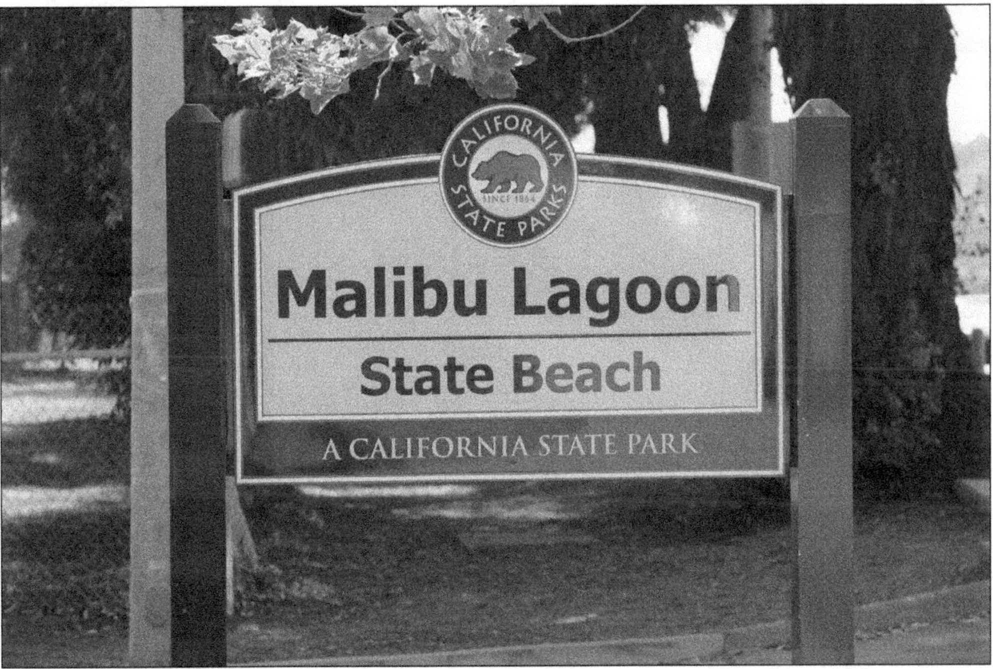

Malibu Lagoon State Beach neighbors the affluent Malibu community. Malibu has over a dozen beaches along the Pacific Coast Highway. The Malibu Cove Colony was popular among Hollywood celebrities dating back to the 1920s. (John Moss.)

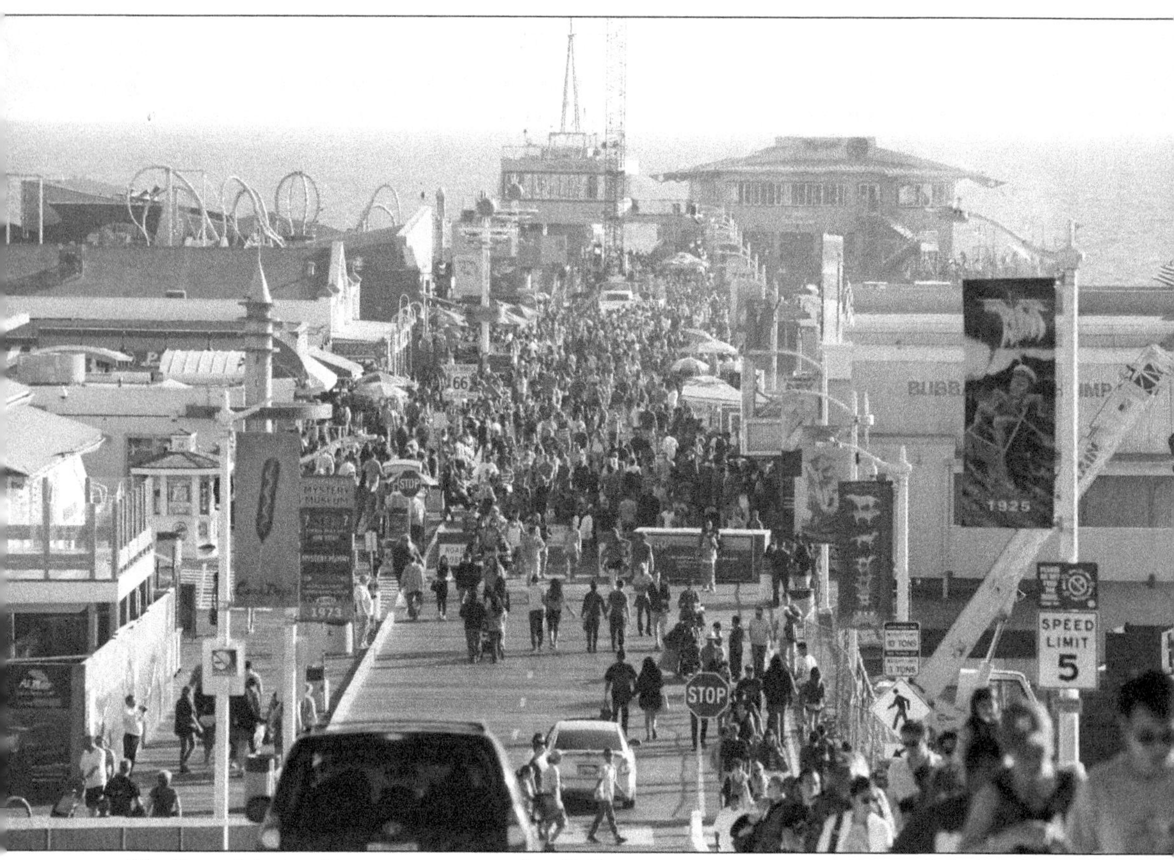

The Santa Monica Pier was constructed in 1909 to transport sewer pipes beyond the breakers. The short, wider pier was constructed as a pleasure pier in 1916. Its landmark is the white hippodrome, which is at the east end of the pier at the entrance. The pier is still operating, and fishing is popular. The park features a carousel, a small roller coaster, and a Ferris wheel. Hundreds of people visit the pier daily, but it is most crowded during the summer months. (John Moss.)

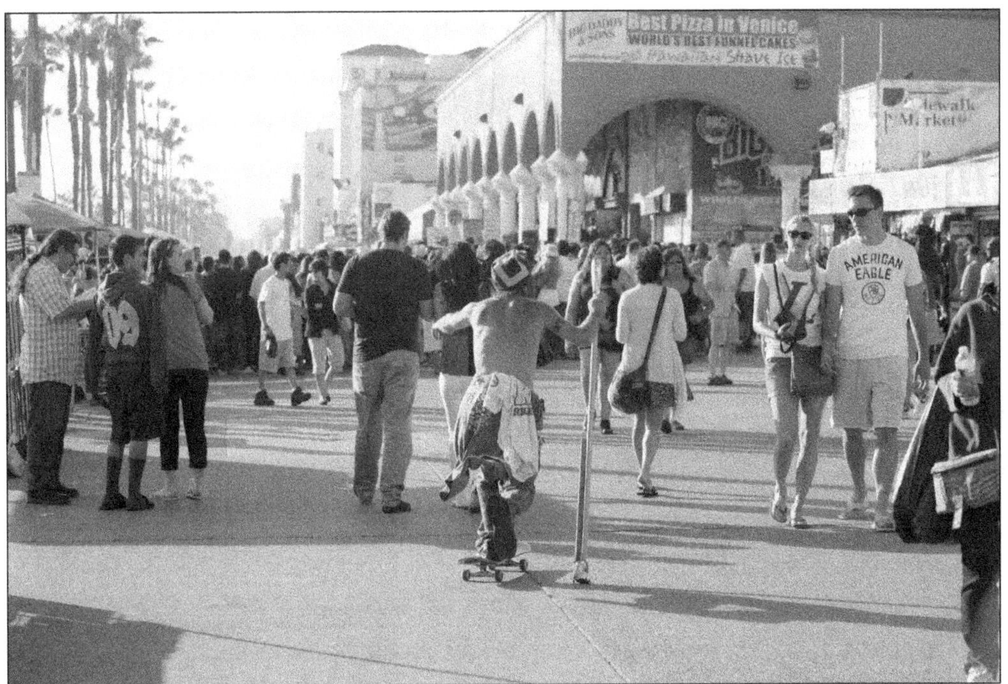

Venice Beach is pictured here as it is seen today. The beach has a boardwalk that runs parallel to the water. The Venice community has its roots in the Ocean Park era, when the community promoted free expression. (John Moss.)

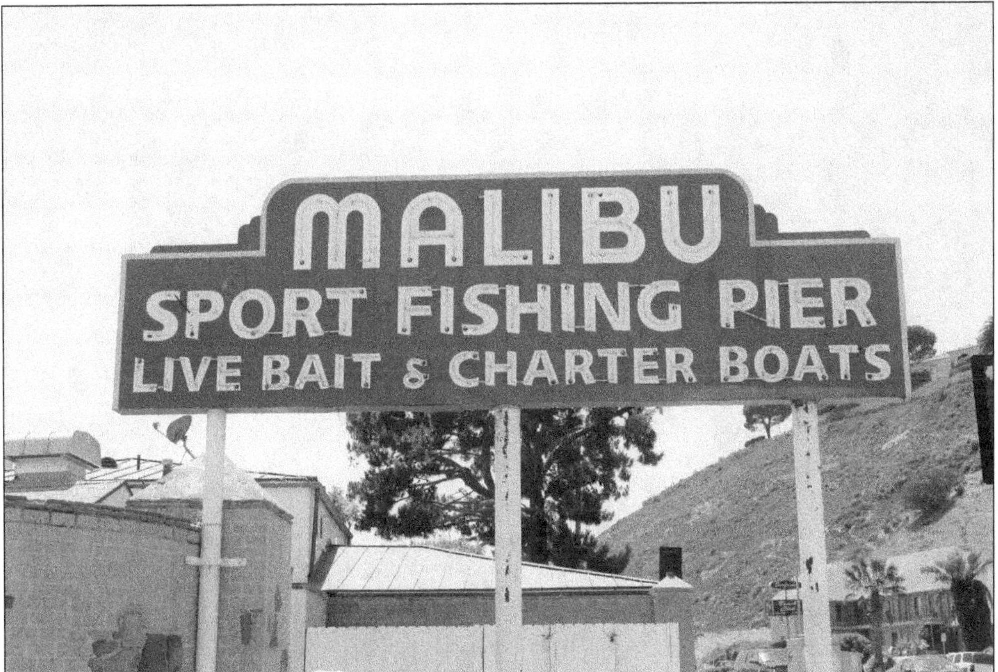

Built in 1905 for the Frederick Hastings Rindge family's business of selling grains, hides, fruit, and other products, this pier also served as a location to transfer goods to ships. It opened to the public for fishing in 1934 and is still in operation today. (John Moss.)

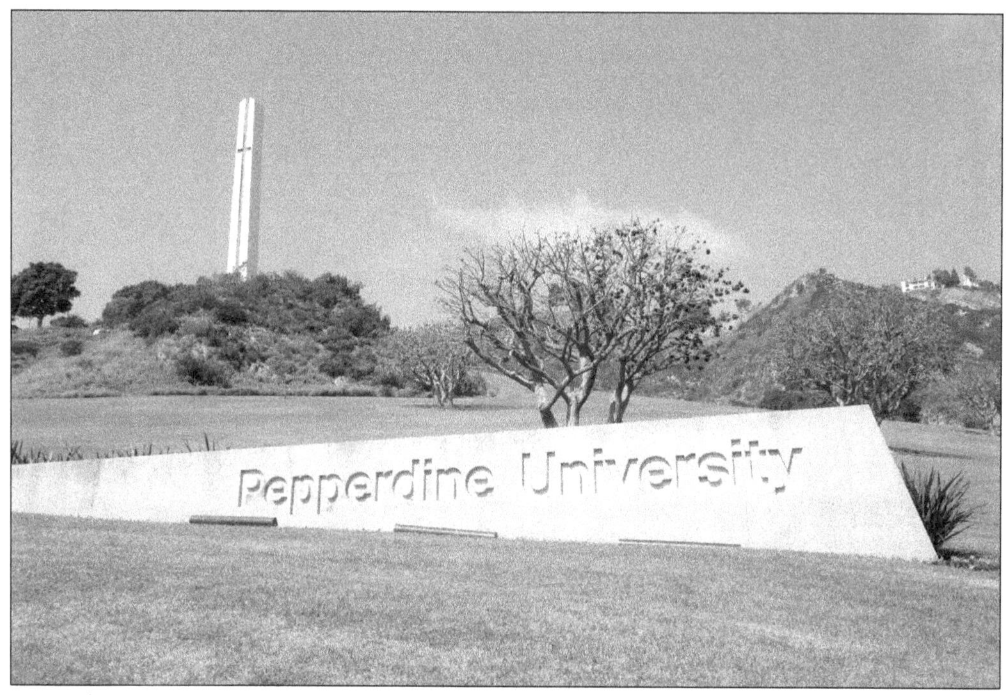

Pepperdine University was originally founded in Los Angeles, but relocated to Malibu along the Pacific Coast Highway in 1968. It broke ground in 1971 and reopened at its new location in 1972. It sits on 130 acres of manicured landscape. (John Moss.)

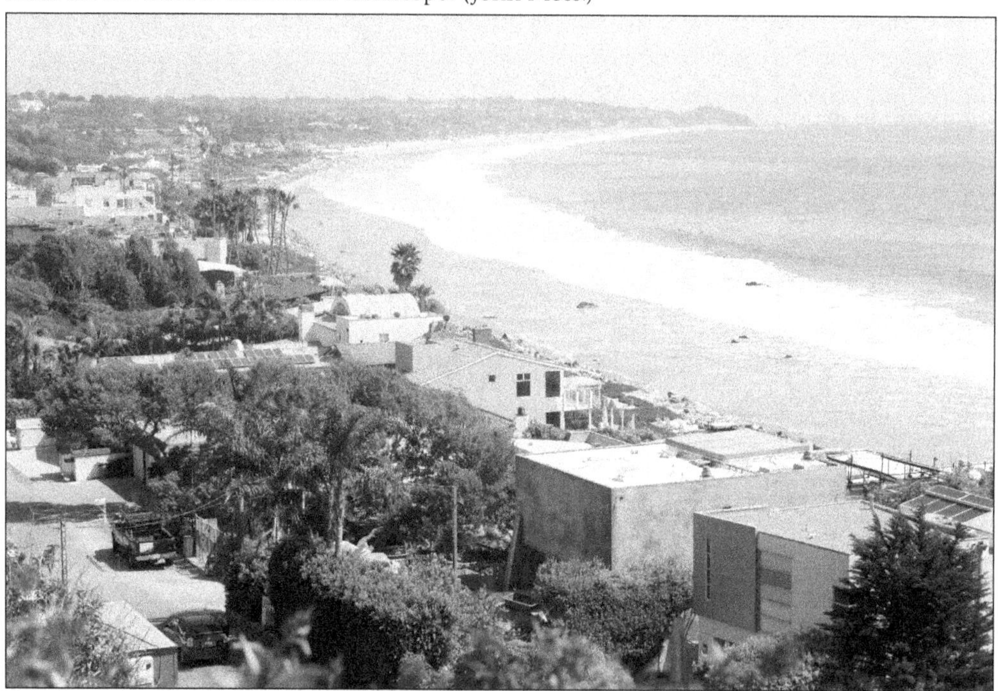

The famed Malibu Colony on the beachfront is still home to many Hollywood celebrities. Celebrities began migrating to the secluded colony in the early 1920s when the Rindge family leased beachfront cottages or lots for tenants to build on. (John Moss.)

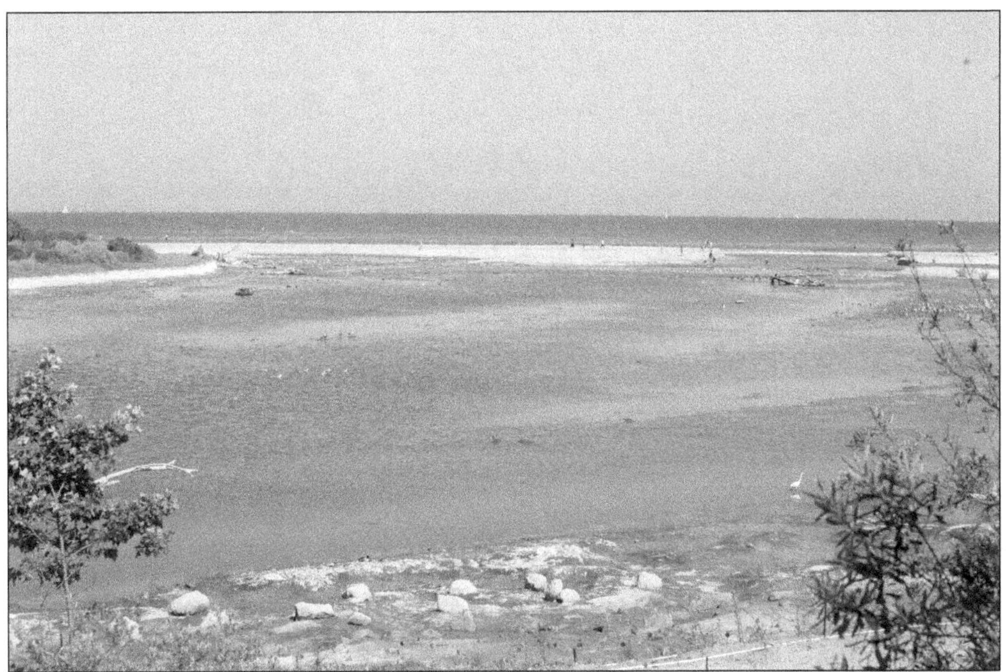

Malibu Lagoon merges with Malibu Creek at the Pacific Ocean. The Adamson House at the lagoon was the home built for Merritt Huntley Adamson and Rhoda Rindge Adamson, daughter of Frederick Hastings Rindge and May Knight Rindge—last owners of the Rancho Malibu Spanish grant. As seen below, waterbirds abound at the lagoon, and a variety come and go with the seasons. Some of the birds include black-legged kittiwake, yellow wagtail, red phalarope, snow goose, and seagulls, to name only a few. It is a favorite place for bird watching and photography. (Both, John Moss.)

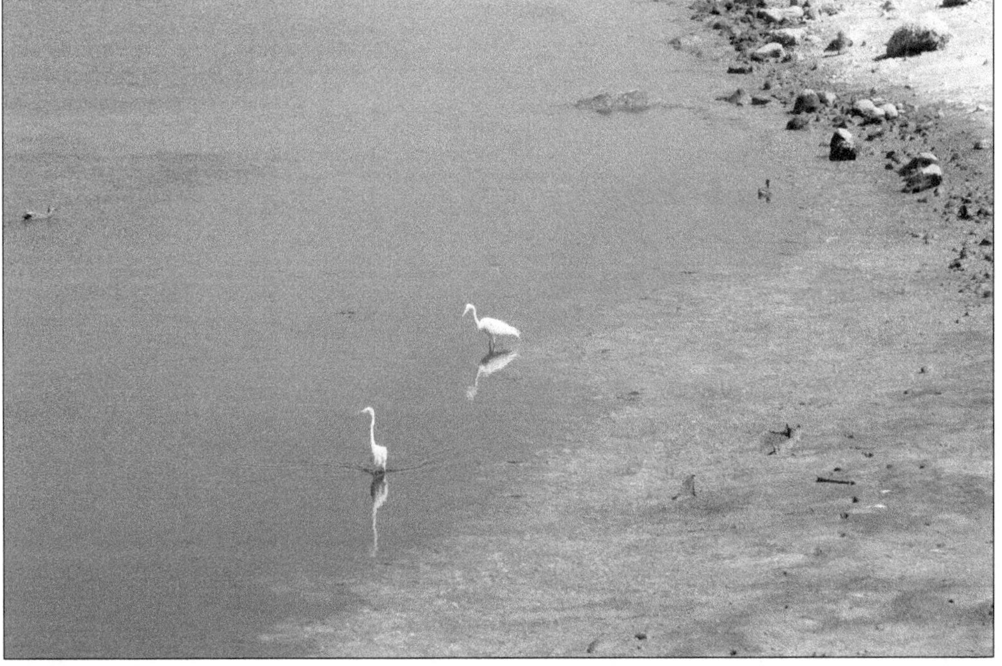

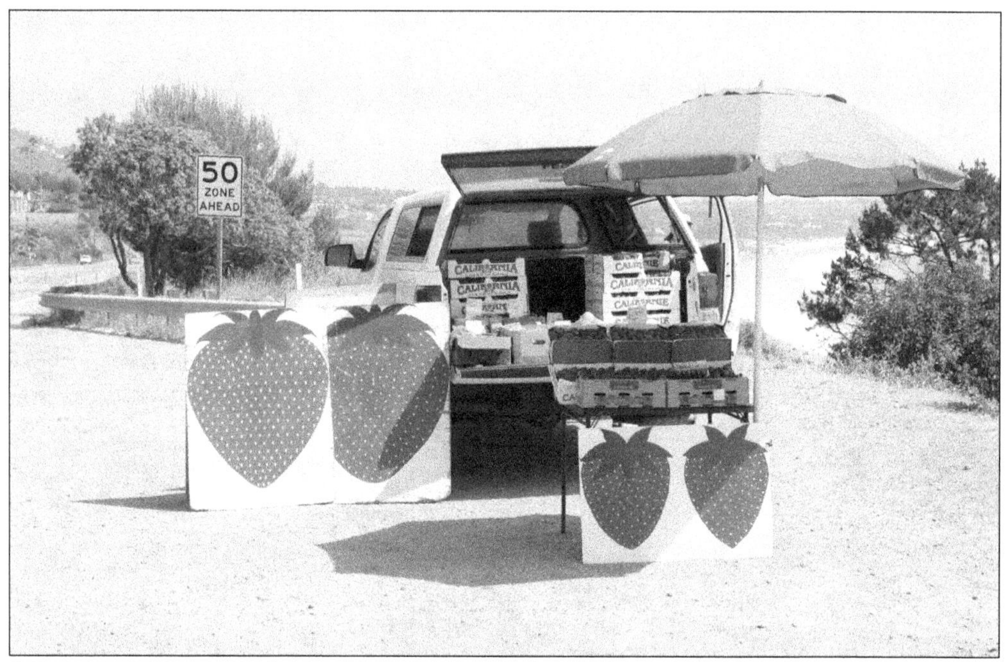

Along Pacific Coast Highway, especially during the spring and summer months in Southern California, there are often vendors parked along turnoffs or on road shoulders selling fruit, seafood, and crafts out of their vehicles. Pictured is a vendor selling strawberries from one of California's nearby strawberry farms. (John Moss.)

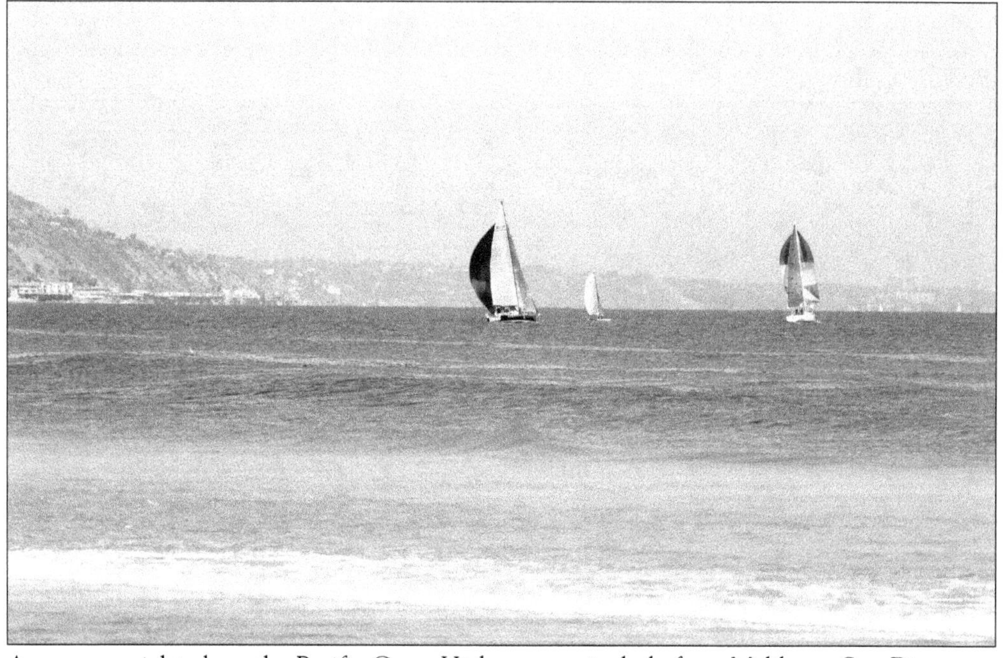

A common sight along the Pacific Coast Highway, particularly from Malibu to San Diego, are the many colorful sailboats and kayaks on the ocean. Southern California has a history of beach and water sports, such as surfing and volleyball tournaments. Many world-famous surfers are from Southern California. (John Moss.)

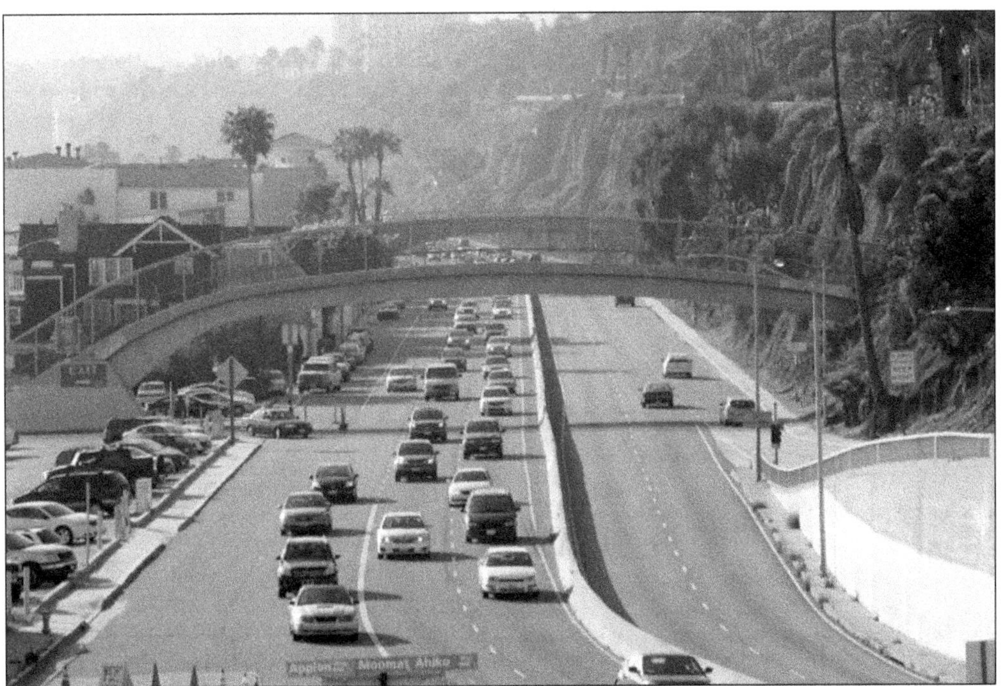

Pictured is a section of the Pacific Coast Highway near the California Incline and Santa Monica Pier, with heavy traffic typical of the area. (John Moss.)

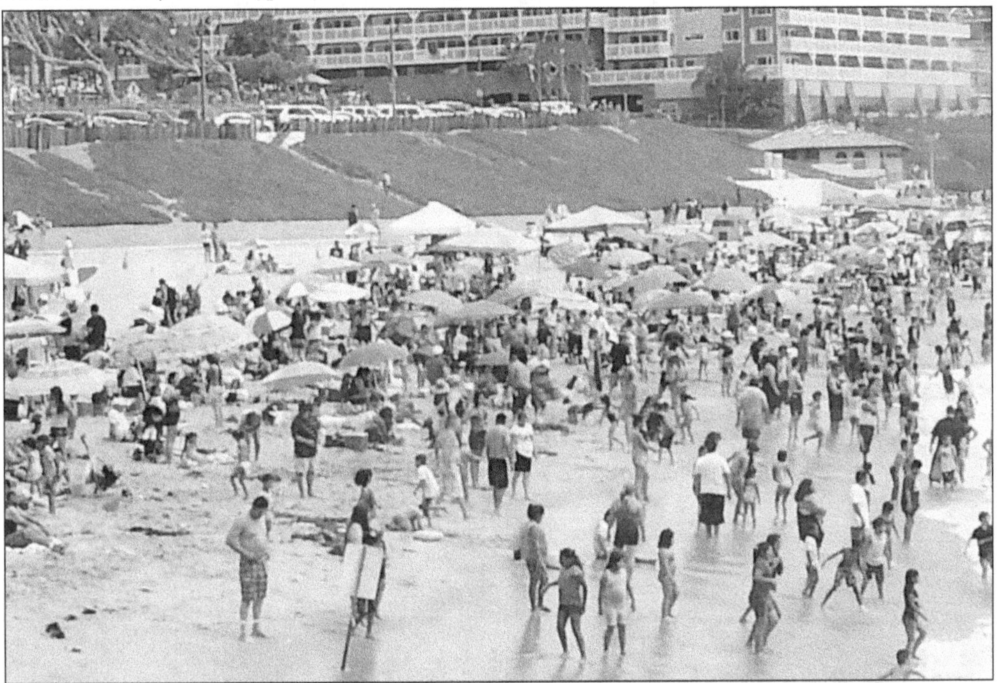

In addition to beachgoers, Redondo Beach colors the ocean with its small boats, paddleboarders, and kayaks. Its King Harbor is a busy pleasure boat harbor. It is a popular beach in the South Bay neighboring Hermosa Beach, another popular spot in Southern California. Surfing competitions have been popular in the South Bay since the 1960s. (John Moss.)

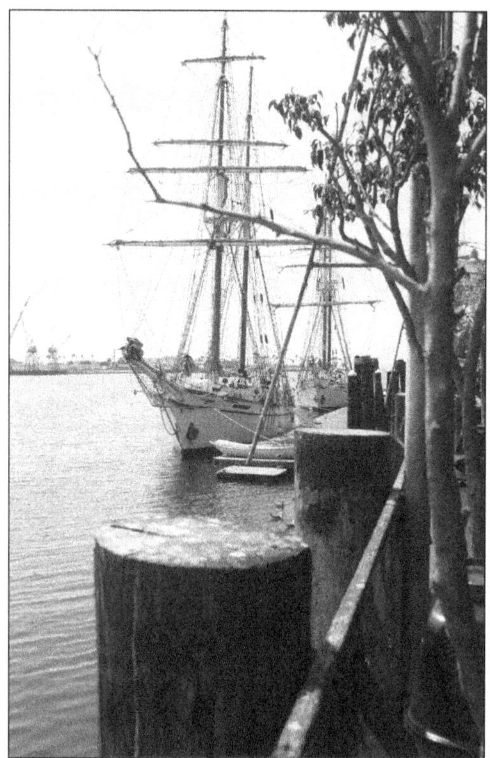

Los Angeles Harbor in San Pedro is the city's main port. Its waterfront includes Ports O' Call Village, where large ships and boats pass. The harbor comprises 43 miles of waterfront—7,500 acres of land and water—and is the largest and busiest container port in the country. (John Moss.)

The historic opening day of the Carmel to San Simeon Highway in 1937, and the dedication of the Elmer Rigdon Memorial Drinking Fountain, marked the birth of a true scenic highway. These two smiling women frame a well-deserved memorial. Dr. John L.D. Roberts's vision became a reality. (Copyright 1937 California Department of Transportation. All rights reserved.)

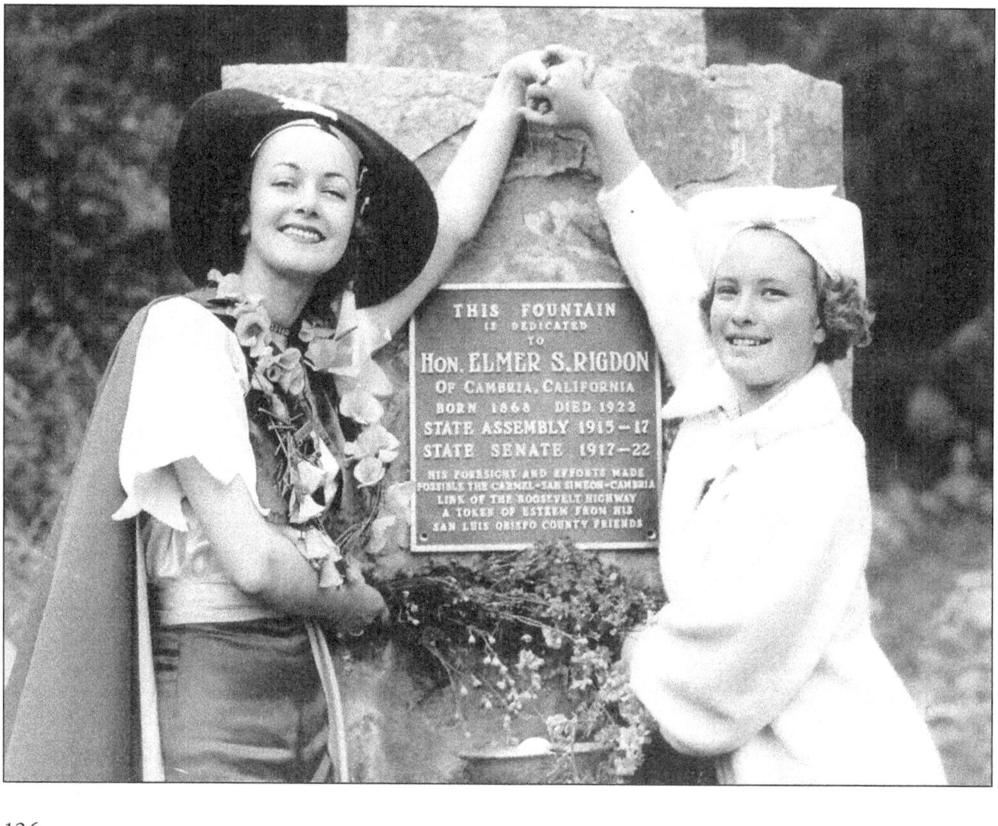

In Southern California, the opening of the Roosevelt Highway in 1929 created a route from Santa Barbara County to San Diego County. When the state widened the highway in 1947, it officially became part of Route 1. (Copyright California Department of Transportation. All rights reserved.)

California's longest north-south highway can take drivers on a journey through some of the oldest and most isolated communities in the state, through majestic redwood forests, over some of the most unique bridges, through tunnels, along narrow twisting roads, and across heights high above the ocean. The only choice is whether to go north or south.

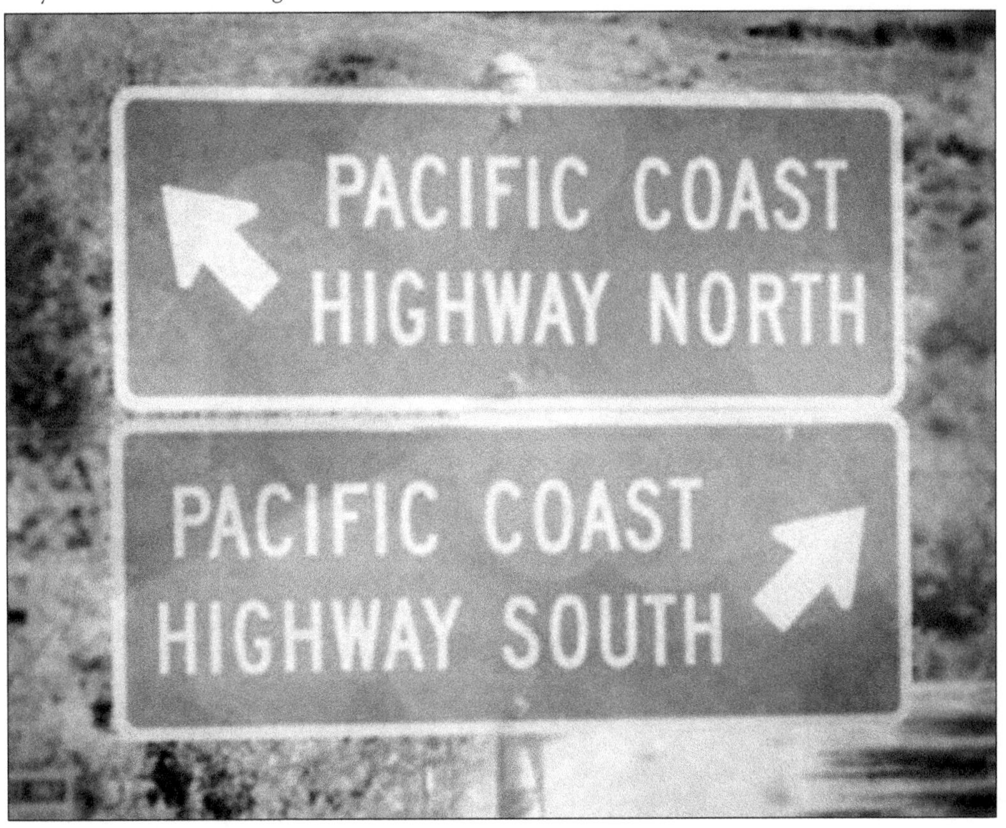

www.ingramcontent.com/pod-product-compliance
Lightning Source LLC
LaVergne TN
LVHW081551100826
845153LV00004B/360
9781540228680